The Accidental Scientist

THE ROLE OF
CHANCE AND LUCK IN
SCIENTIFIC DISCOVERY

Graeme Donald

Michael O'Mara Books Limited

First published in Great Britain in 2013 by
Michael O'Mara Books Limited
9 Lion Yard
Tremadoc Road
London SW4 7NQ

A CIP catalogue record for this book is available from
the British Library.

Papers used by Michael O'Mara Books Limited are natural,
recyclable products made from wood grown in sustainable forests.
The manufacturing processes conform to the environmental
regulations of the country of origin.

ISBN: 978-1-78243-015-5 in hardback print format
ISBN: 978-1-78243-099-5 in EPub format
ISBN: 978-1-78243-100-8 in Mobipocket format

1 3 5 7 9 10 8 6 4 2

Cover design by Ana Bjezancevic
Designed and typeset by Envy Design Ltd
Printed and bound by CPI Group (UK) Ltd, Croydon, CR0 4YY
Picture research by Judith Palmer

www.mombooks.com

For Rhona, my own personal gift from
the Princes of Serendip

Contents

Introduction

ACKNOWLEDGED AS ONE of the most difficult words to translate into any other language, serendipity – as a word and a concept – was invented by Horace Walpole, son of Robert Walpole (who is broadly recognized as Britain's first Prime Minister, despite that office not existing until 1937 – all previous incumbents being formally titled First Lord of the Treasury; but we digress). Horace was inspired by the ancient tale of *The Three Princes of Serendip* – the old name of Sri Lanka – which tells of many discoveries and situations successfully resolved by chance or blunder, as indeed has been the case throughout the history of science and medicine.

Take, for example, the discovery of PTFE, better known today as Teflon. Although non-stick pans are cynically and wrongly said to be the only benefit accorded the general population by the space programme, the substance was in fact discovered by chance in 1938 by Du Pont's Roy Plunkett while he was working on refrigerants. A cylinder of tetrafluoroethylene gas failed to discharge, despite the fact that its weight indicated that it was full. At this point most would have simply grabbed another cylinder, but not Plunkett, who cut the cylinder in half to see what was going on. The inside was coated with a white deposit indicating the gas to have polymerized. This white deposit has meant omelettes have been easier to cook ever since.

Sticking with the American space programme, serendipity can also work in reverse, chronologically speaking. In 1962, NASA was struggling to design the spacesuits that, in 1969, would be worn by the first men on the moon. In a chance conversation, one of the team was correcting a colleague who had trotted out the old myth about armour being so heavy that knights had to be hoisted onto horses by small cranes. While lecturing his teammates on the lightness and flexibility of such suits that in fact rarely weighed more than 50lb, it dawned on all present that the answer to their problem might lie in the past. The team flew hot-foot to the UK to visit the armouries at the Tower of London and subsequently modelled their famous moon attire on a suit of armour made for Henry VIII to fight on foot in knightly contests. The secret lay in the articulation of all the joints, which allowed for full radial movement. Today, visitors to the Tower can still see the moonsuit sent from America in gratitude, standing aside its historical inspiration.

There are, of course, many more examples of serendipitous discovery than those included in the following pages – even the recreation drug ecstasy evolved from a 1953 search for a truth-drug conducted by the US Army – but everyone involved in the project hopes you enjoy the stories laid out in *The Accidental Scientist* and will perhaps be encouraged to seek out more examples. (Actually, I wanted to call it *Serendipity-Do-Dah* but they put the block on that pretty smartish.)

Botox

TOXINS CAN BE FUNNY THINGS; some, like snake venom, which is little more than a modified enzyme that pre-digests the snake's prey, are lethal if administered intravenously yet can be ingested without harm. Others, like botulism, can be lethal if ingested yet benign or even beneficial if injected – under the right conditions.

Usually associated in the general mind with contaminated meat, *Clostridium botulinum* is also pandemic in the soil and finds low-acid vegetables, such as asparagus, an ideal host; most dangerous of all is the humble baked potato if wrapped in baking foil after cooking and then left at ambient temperature. The first to suspect the existence of this toxin was the German poet/physician Justinus Kerner

(1786–1862) who, in his home town of Württemberg in 1817, traced an outbreak of such food poisoning to a batch of boiled sausage and so named the culprit from the Latin *botulus*, a 'sausage'. Such outbreaks were far more common in Württemberg than in other towns and cities, an anomaly Kerner speculated might be due to the local habit of slow and low temperature boiling to reduce the likelihood of the sausage bursting; furthermore, although he had no idea as to the nature of the agent, he was the first to speculate on possible medical uses. But it would be a chance invitation to, of all things, a funeral, seventy-eight years later, that cracked the matter.

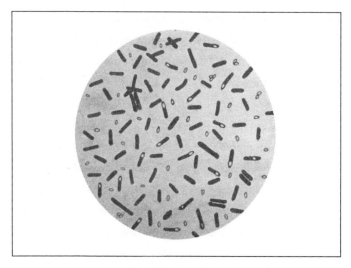

Clostridium botulinum

OOM-PAH

On 14 December 1895, the Belgian microbiologist Emile van Ermengem (1851–1932) was attending the funeral of one Antoine Creteur, in the farming community of Ellezelles where he was living at the time. Present at the post-interment shindig was the Fanfare Les Amis Reunis, a still-extant brass band famed throughout modern Belgium. After their efforts, the band members took themselves off to Le Rustic tavern for beer and a snack of the locally produced air-dried and smoked ham; a speciality not unlike Parma ham. Soon, most of the thirty-four band members started presenting symptoms of blurred vision, flaccid muscles and slurred speech, a not uncommon experience after a long session in a pub, but then they started dying. Three of the youngest musicians were the first to succumb: Jules Hautru and Angel Deltenre, both nineteen-year-old farm labourers, and a twenty-two-year-old saddler Firmin Cretuer, a relative of the recently interred.

The crucial factor, of course, was that the handful of musicians who had shunned the ham in favour of other foods remained fit and well, this giving Ermengem a God-given opportunity to immediately identify the source and set to work.

He returned to his laboratory in the University of Ghent with samples of the ham, which was liquidized and injected, fed to or implanted under the skin of assorted rabbits, dogs and monkeys, all of which took rapid onset of identical symptoms and died. Within a matter of weeks he had isolated and identified the bacterium responsible and published papers on its nature and, more importantly, how to eliminate its growth in the air-dried ham process.

MONKEY BUSINESS

In 1946, Dr Edward J. Schantz succeeded in producing a crystalline form of the toxin which allowed its study in greater detail. He was followed in the 1950s by Dr Vernon Brooks who, experimenting with the toxin on a group of monkeys, one of which just happened to have a rather pronounced tic, noticed that this abated after each injection. It was soon established that the toxin was blocking the release of acetylcholine from the motor nerves that were driving said muscles into spasm. In turn, this fortuitous side effect, mentioned in Brooks' papers, inspired Dr Alan B. Scott of San Francisco's Smith-Kettlewell Eye Research Foundation to see if the same toxin might relax muscles causing strabismus, or cross-eyes. Finally, and with FDA clearance in 1978, injections of botulism became standard treatment not only for strabismus but also for facial and neck spasms and assorted nervous tics, all of which disappeared over night.

DOES BOTOX MAKE YOU HAPPY?

Some recent research claims to have established a link between Botox injection to the face and a lack of ability of the injectee to empathize with the emotional plight of others, or at least this is the claim of David Neal, Professor of Psychology at the University of Southern California. A similar conclusion is reached by Barnard College professors Joshua Davis and Ann Senghas with Davis opining:

> With Botox, a person can respond otherwise normally to an emotional event, e.g. a sad movie scene, but will have less movement in the facial muscles that have been injected, and therefore less feedback about such facial expressivity . . . It thus allows for a test of whether facial expressions and the sensory feedback from them to the brain can influence our emotions.

Or, it might just be that the narcissistically self-obsessed simply don't care about others' problems and never did. That makes much more sense and is a more speedily achieved conclusion.

SMOOTHING THINGS OVER

But the cosmetic stampede would not begin until 1987 with a few chance comments from the patients of husband-and-wife team Drs Jean and Alastair Carruthers of Vancouver. Jean was using botulism injections in their joint practice where most of her patients were afflicted with facial spasm or blepharospasm, an uncontrollable blinking and screwing up of the eyes. One of her female patients, whose forehead muscles had ceased to spasm, returned demanding another injection anyway because her wrinkles had come back. Jean mentioned this to her husband over dinner that night and he sat up attentively. He had for some time been on the edge of cosmetic work and using various fillers on patients' glabellar lines, the vertical frown-lines between the eyebrows.

The next day at the surgery they badgered their thirty-year-old receptionist Cathy Bickerton Swann into becoming the first human guinea pig for cosmetic reasons alone. Cathy had developed some rather pronounced furrows in her forehead, these descending in a V-shape to a point between her eyes, giving her a slightly 'Klingon' appearance that, if she actually *did* frown on top of all that, made her look 'pretty damned hostile', according to the Carruthers. The results were pronounced to say the least; smooth as the proverbial, according to all three, and within days the practice was besieged by a Barbie-queue trailing half-way round the block, all of them begging for injections of what would soon be marketed as Botox.

As for Cathy, she continued the injection for a while, only to please the Carrutherses. But she left their employ soon after and, at her last interview, proclaimed herself to have been Botox-free ever since: still lined and a bit on the portly side, but extremely happy with her life and her role in it.

DNA Fingerprinting

WHEN SIR WILLIAM HERSCHEL (1833–1917), son of the famous astronomer, forced a contractor to leave his inky handprint on a contract, he had no idea of the future implications for criminal forensics and identification. Completely unaware of the unique features of the mass of lines and whorls, Herschel, then of the Indian Civil Service at Jangipur, was fed up of being cheated by local contractors paid in advance who either reneged on the conditions, saying they had never seen the paperwork in the first place, or sent along a family 'lookalike' to pretend to be them and swear blind it had not been them in the first place.

Herschel had no idea that the handprint could *actually* be used for identification, he only wanted to scare contractor Rajyadhar Konai into thinking he could be identified by the print at some later date if he failed to fulfil the road-building contract on time. It worked; the wily Konai, who had weaselled his way out of previous commitments, placed his inked hand on the contract on 28 July 1858 and subsequently complied with the terms and stipulations of that document. Chuckling to himself at the native gullibility, Herschel expanded the use of what he thought was a pointless ruse, gradually reducing the required print to just the tips of the fingers of the right hand.

THE SWAP SHOP

In 1877, Herschel took over as Magistrate of Hooghly and thus became responsible for not only the criminal courts but also the Registration of Deeds. Still blissfully unaware that he was actually on to something quite groundbreaking, Herschel extended the practice to the prison system and the Army Pensions issue, to eliminate the double-payment fraud that was a serious drain on resources. In the case of the local prison, Herschel was determined to cut down on the local habit of families of the inmates giving their nearest and dearest a holiday from the rigours of hard labour. Again using lookalike surrogates, the family would send along a visitor to change places with the convict for an agreed time, after which the real miscreant would return to reverse the ploy. The guards didn't really mind, as long as the head-count was right, but

Herschel thought this unsporting and insisted on all visitors to be fingerprinted on the way in and out of the jail.

And it worked; Herschel, and in fact the whole of the Western world, may have been unaware of the validity of this 'new' measure but perhaps the locals knew better, because it has been subsequently established that the uniqueness of a person's fingerprints was well known throughout the Eastern world. Later research reveals that the Chinese were collecting hand, foot and fingerprints from crime scenes as early as the year 300 and using them in subsequent court proceedings, as indeed were the Japanese and Iranians. The earliest reference to fingerprints' uniqueness so far discovered is to be found in Rashid-al-Din Hamadani's (1247–1318) *Universal History* in which the author observes: 'Experience shows that no two individuals have fingers exactly alike.'

OUT OF PRINT

The first Westerner to publically advocate the use of fingerprints in criminal identification and detection was the Scottish missionary-physician Henry Faulds (1843–1930) who had based himself at Tokyo's Tsukiji Hospital. In the summer of 1880 the visiting American orientalist Edward S. Morse (1838–1925) invited Faulds to accompany him on an archaeological dig. Pressganged into spending his free time trudging up-country, Faulds was struck by the fingerprints of ancient potters still clearly visible in the fragments unearthed. More striking was the fact that the native excavators seemed to be matching up the fragments, wherever possible, by the fingerprints of those same ancient potters.

Now immensely pleased that he had accepted Morse's invitation, Faulds started a fingerprint collection of friends and colleagues back at the hospital, just in time to save one such colleague from false arrest. With Faulds in the middle of gathering prints and data to back up his notion that each person's print was a unique form of identification, the hospital was broken into and a colleague arrested by the local police. Faulds gathered fingerprints from the crime scene and, after presenting his evidence, secured the release of the colleague. He made reference to this vindication in a paper published on 28 October 1880:

> When bloody finger-marks or impressions on clay or glass exist they may lead to the scientific investigations of criminals. Already I have had experience of two such cases and found useful evidence from these marks. In one case the greasy finger-marks revealed who had been drinking some rectified spirit. The pattern was unique and fortunately I had previously obtained a copy of it. They agreed with microscopic fidelity. In another case sooty finger-marks of a person climbing a white wall were of great use as negative evidence.

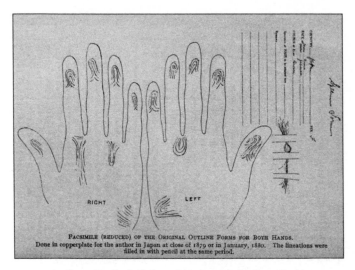

FACSIMILE (REDUCED) OF THE ORIGINAL OUTLINE FORMS FOR BOTH HANDS.
Done in copperplate for the author in Japan at close of 1879 or in January, 1880. The lineations were filled in with pencil at the same period.

An illustrative facsimile from *Dactylography* by Henry Faulds

Faulds was destined to fall out with everybody as a result of this publication. Trying to gain some heavyweight support to the use of fingerprinting in the fight against crime he wrote to Charles Darwin who deciding not to get involved passed the letter on to his cousin, Francis Galton, the chap who came up with the dark notion of Eugenics. When Galton realized that there was nothing in fingerprints to help identify criminal types and deviants *per se*, and so weed them out of his dream of a genetically cleansed society, he too lost all interest for the time being.

SMUDGED PRINTS

But how reliable is fingerprinting evidence; is it the infallible forensic factor that most imagine? While each individual's fingerprints have, so far, been proved unique, as with DNA fingerprinting, the problem lies in good old-fashioned human error and the reading of the evidence. Prints at crimes scenes are rarely found in neat array on flat surfaces, they can be partial, smudged or, in the case of, say, a ball of putty or plastic explosive, distorted, elongated or otherwise 'scrunched up'. The most celebrated such false identification was that of Brandon Mayfield (b. 1966), a lawyer of Portland Oregon who, with his fingerprints on file for some juvenile misdemeanour, was arrested on 6 May 2004 for his alleged involvement in the Madrid train bombings. Evidence that he had not even left the USA for over fifteen years was swept aside by the fingerprint identification and he was held incommunicado, with no right to attorney consultation, for over two weeks until the mix-up was sorted. Mayfield became something of a *cause célèbre* and focused international attention on the writings of Dr Simon Cole, Professor of Criminology at America's Cornell University who has for years cautioned against courts' blind reliance on fingerprint evidence; by his reckoning, there are about 1,000 such false convictions per year in the United States alone.

CRIME SCENE INVESTIGATION

And then came the first salvo from William Herschel who, miffed that he had missed the significance of his hand and fingerprinting for all those years, waded into the fray with claims to have beaten Faulds to the concept by nearly thirty years. The two men entered into a bitter war of public and private exchanges that would endure until Herschel's death. Not to be left out, the French published translations of the writings of their most extraordinary master criminal-turned-detective and forensic criminologist, Eugène Vidocq (1775–1857).

The French authorities, desperate to combat the crimewave for which Vidocq and his associates were largely responsible, colluded in his escape from prison in 1809 on the understanding that he would repay by setting up a covert thief-catching unit. The authorities were so impressed with Vidocq's clear-up rate that his unit evolved into the still-extant Sureté Nationale. The British were, of course, outwardly shocked at the very thought of dealing with the Vidocqs of this world, but, on the quiet, they sent over a fact-finding team to study Vidocq's methods to help Sir Robert Peel more properly structure his new venture at Scotland Yard. Vidocq was not only the first to take shoeprints from crime scenes and match them to criminals' footwear but he was also, as early as 1820, trying to take fingerprints for the Sureté files – only the type of ink foiled him. Using ordinary ink that dried too quickly to be of much use, Vidocq would have been years ahead of the game had he only thought of using the kind of slow-drying and oil-based favoured by printers. To be fair to Herschel, he had been using the ink for his official stamp and Seal of Office, which was pretty much the same as printers' ink, so perhaps he can claim a bit of the glory.

FINGERING IT OUT

Eight years after receiving Faulds's letter, Galton looked again at fingerprints and published a paper detailing their broad categorization and definition, which helped greatly with their adoption by police forces around the world. He didn't mention Faulds's letter, which left him open to a withering barrage of letters from the increasingly marginalized old Scot who, quite rightly, felt robbed.

The first police force to adopt the Galton Fingerprint Classification System, as it became known, was the La Plata Division of Buenos Aires, which secured the first criminal conviction from fingerprint evidence alone. On the 29 June

1892, Francesca Rojas ran bloody and screaming from her house in Necochea, claiming to have been attacked and her children murdered by her neighbour, Pedro Velasquez. But with Pedro sticking to his story, no matter how much they 'interrogated' him and the emerging knowledge that Francesca had a lover who refused to marry her because of the children, the police looked again at the house and found fingerprint evidence pointing at the mother herself. The first conviction in the UK was somewhat less dramatic: Harry Jackson climbed through an open window of a house in London's Denmark Hill and stole a set of billiard balls; the window was freshly painted and he left his thumbprint for all to see.

FUMING MARVELLOUS

The next accidental step taken in fingerprinting happened in 1977 at the National Crime Laboratories in Japan when fibre expert Fuseo Matsumur was working on trace evidence gathered at the scene of the murder of a taxi driver. Securing the samples to slides with superglue he popped them under the microscope and, to his surprise, saw his own fingerprints slowly appear on the reverse of the first slide. After consultation with his colleague, Masato Soba, and a few more experiments with their own fingerprints, they realized that cyanoacylate fumes from the glue were highly attracted to the kind of amino and fatty acids left by latent prints that were otherwise invisible. Deciding that the fumes were the major benefit they knocked up a glass box with an impromptu heater and the results were quite staggering. Today, superglue-fuming is standard practice around the world.

EUREKA!

And finally, the happy accident of DNA fingerprinting. Sir Alec Jeffreys (b. 1950) was tinkering with X-rays of samples of genetic material harvested from his associates in the Department of Genetics at Leicester University when, at 9.05 on the morning of 10 September 1984 – Jeffreys is nothing if not meticulous when it comes to his notes and records – he had his eureka moment. Developing an X-ray that had picked up random DNA from a technician and her mother and father, Jeffreys looked at the 'blob' and recalled:

It was nothing to do whatsoever with forensics. It was basically aimed at human genetics and medical genetics . . . Five minutes before that first fingerprint I didn't have a single thought about forensics at all . . . I took one look, thought 'what a complicated mess', then suddenly realized we had patterns . . . There was a level of individual specificity that was light years beyond anything that had been seen before . . . It was a 'eureka!' moment. Standing in front of this picture in the darkroom, my life took a complete turn. We could immediately see the potential for forensic investigations and paternity, and my wife pointed out that very evening that it could be used to resolve immigration disputes by clarifying family relationships.

Within a matter of months the 'reading' process was refined to such a point that the DNA fingerprint was clearly defined but, today, there are growing concerns over police reliance on

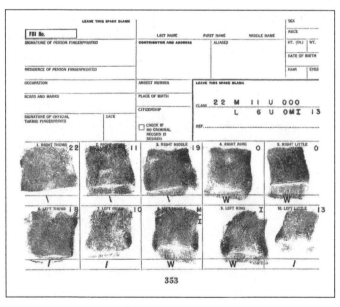

Fingerprint samples taken by the FBI in the 1960s

even this method of identification. At the time of writing only ten markers of similarity are required to produce what the courts deem a positive match which some feel should be raised to fifteen or more. Also, that juries should be instructed that DNA evidence only speaks in terms of statistical likelihood and that samples gathered can, quite inadvertently, be subject to cross-contamination in the lab or even maliciously placed at the scene of the crime by the real perpetrator. All you now need to frame someone for murder is, say, a few hairs from their hairbrush, a cigarette they have smoked and discarded, a glass they have drunk from or a used tissue from their bedroom bin. Makes you think, doesn't it?

Cellulose

THE STORY BEHIND CELLULOSE is riddled with chance and accidental discovery but readers should bear in mind that it was, and still remains, pretty dangerous stuff.

In 1832, the French chemist Henri Braconnot (1780–1855) established that wood fibres treated with nitric acid and allowed to dry resulted in a fun but rather unstable explosive. A few years later, the French chemist Théophile-Jules Pelouze (1807–67), who would later put both Alfred Nobel and Ascanio Sobrero on their respective paths to glory, likewise treated paper and cardboard to produce flash-paper to amuse children and street urchins. But the man who put nitrocellulose on the map was the Swiss–German Christian Friedrich Schönbein (1799–1868).

YOU SMELL GAS?

Chance had already smiled on Christian in 1840 when he noticed a strange smell while walking through a laboratory in the University of Basel, where experiments were being conducted on the electrolysis of water. The smell reminded him of the detectable odour in the air during a thunderstorm. Rightly deducing there to be a new gas on the block, he christened it ozone from the Greek '*ozien*', to smell.

COOKING UP A STORM

In 1844, Frau Schönbein was suddenly called away to a sick relative. With his wife out of sight, Christian started to turn her kitchen into a laboratory and soon spilled a mixture of nitric and sulphuric acid on her table. Reaching out in panic for the first available thing, he mopped up the acid before it could do any real damage. He then realized that he had in his hand her apron, which he had blindly snatched from its peg by the stove. Knowing better than to soak it in water, he carefully returned it to its peg to dry. Checking it later for stains, he was relieved to see that the apron appeared none the worse for its adventure so he gave it a jaunty flick to smarten it up before returning it to its perch. When he came round on the

A 'Kneading, Mixing, & Incorporating Machine' for smokeless and flameless gunpowders, manufactured in London

other side of the kitchen, surrounded by concerned neighbours, he realized he had invented gun-cotton, cotton itself being naturally high in cellulose content. This became the smokeless powder that would allow armies to kill each other in much higher numbers, as they would now be able to see each other instead of each standing in a thick pall of their own gun smoke.

BURN CARD

No matter how dangerous in the right circumstances, nitrocellulose permeates modern life; it is the Sellotape at the office and the gunk that holds together bars of staples – but it is its presence in playing cards that presents the greatest danger to those on the wrong side of the law. In October 1930 William Kogut languished on San Quentin's death row for the murder of brothel-keeper Mayme Guthrie. Not wishing to meet Old Sparky he cut up some playing cards into tiny pieces, soaked them in water and, along with a makeshift bullet, packed them into the blind end of a steel leg torn off his bed. He placed the 'loaded' end on the cell's heater and the open end against the side of his head and waited; it worked just fine.

More recently, the so-called Birmingham Six invoked the playing card rule. Known supporters of the Irish Republican cause, they were convicted in 1975 of the murder of those who died in the bombing of two pubs in the New Street Station complex, Birmingham. Shortly before the explosions, the six men boarded the ferry-train from that same complex to attend the Belfast funeral of an IRA bomber. Forensic nitrate tests indicated that two of them had been handling explosives, but this was later called into question by the fact that they had been playing cards on the train before their arrest at Heysham ferry-port and that merely shuffling a pack would be enough for a positive Griess Test, as it is known.

EXPLODING BALLS

Twenty years on from Schönbein's experiment, in 1863, the increasing demand for billiard balls had driven the African elephant to the brink of extinction – by that time, 12,000 animals were being killed every year for billiard balls alone.

Alarmed by the rising cost of ivory, the American billiard company of Phelan and Collender offered a $10,000 prize to any bright spark who could come up with a viable synthetic alternative and, with his eye on the money, a twenty-six-year-old New Jersey printer called John Wesley Hyatt (1837–1920) stepped up to the plate. Experimenting and rejecting various compounds, Hyatt cut himself while chopping up paper to mix into a fresh batch, so he went to the cupboard for the bottle of collodion kept on site for just such eventualities. 'Collodion' is basically nitrocellulose dissolved in alcohol, which was then used to dress wounds – spray-on dressings are nothing new and still contain much the same ingredients. At first, Hyatt was dismayed to see that the bottle had been knocked over some time ago and, with all the alcohol evaporated, the base of the cabinet was sheathed in a hard sheet of cellulose nitrate. But Hyatt had found the material for his billiard balls.

With his brother, Isaiah, Hyatt scooped the prize and set up a new business making billiard balls, and a host of other products, to become wealthy men. But there were one or two drawbacks with the products and all these down to our old friend nitrocellulose hiding in the mix. Billiard balls, by their very nature, are required to bang together, and the brothers soon received a letter from a Colorado saloon-keeper who was in some distress at the number of times Hyatt's balls exploded during play, this in turn prompting some of his more 'frontier-minded' customers to draw pistols and stand ready to shoot. Celluloid-based shirt-fronts, aka dickies, were very popular for their enduring whiteness and water-resistance but there was the odd accident with hot ash falling from cigars. Mothers, used to chastising wayward offspring with their old-fashioned wooden hairbrushes continued to do so with their bright and shiny new celluloid ones, sometimes giving little Johnny a great deal more to think about than perhaps he deserved. And then there were the false teeth . . .

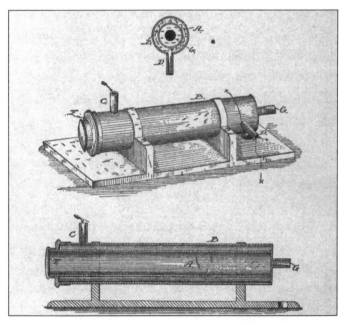

The Hyatt Gun Method of celluloid billiard ball manufacture

A WHALE OF A TIME

London-born Robert Chesebrough (1837–1933) was a young chemist who ran a factory making fuels from oil harvested from sperm whales when the Oil Rush of 1850s America put him out of business. The first strike had been made in Titusville, Pennsylvania so, reasoning that was the place to be, he booked passage on the first available ship for America to see if he could at least get a job there and learn about the new business. The first thing he noticed was that any driller or rigger who injured himself would cover the wound with some of the thick gunk that always came up on the drilling rods when they were extracted from the ground. Furthermore, he noted that the wounds and burns did indeed seem to heal much faster than if left unattended.

Fascinated, he gathered up samples of the gunk and set up a workshop to refine the gel and make it lighter in colour and purify it to be more pleasing to the eye without diminishing its curative effects. His first attempt at marketing Rod Oil was not a success so he came up with the rather more acceptable name of Vaseline, a compound of the German for water and the Greek for oil.

DANGEROUS DENTURES

Those sporting the shiny new gnashers soon realized that their celluloid dentures were quite sensitive to heat and had a tendency to leave the wearer looking a trifle vampiric if the coffee was drunk while still hot enough to melt the teeth into a decidedly pointed profile, but that was the least of wearers' problems. A few unfortunate chaps, left tired and emotional by an excess of bourbon and thus confused as to which end of the lighted cigar should go in the mouth, had nearly blown their heads off. *The Herald* newspaper of Carroll, Iowa, dated 27 May 1908, carried the cautionary tale of the fate of Moorehead resident Cyrus Hopkins who, sporting his new teeth, paused in the bar of Rogers Hotel to light a cigar and instead set fire to his ZZ Top-like beard. Having previously survived an exploding denture incident, Cyrus had the presence of mind to keep his mouth tight shut and, instead of screaming for help, stood there flailing his arms and rolling his eyes to attract the attention of hotel clerk, William McGuinn, who, with a well-aimed glass of beer, and a bar-towel follow-up, extinguished his now almost clean-shaven customer. It was said in the America of the day, that if you listened very carefully you could hear the elephants laughing.

Synthetic Dyes

BEFORE THE ADVENT of synthetic dyes, people had to be content with whatever colorants they could extract from nature. Typically, this came down to whatever was locally available so you could, in such times, make an educated guess as to where strangers hailed from by the colour of their clothes. It was this locality of source that gave rise to the variety of tartans in early Scotland and not, as myth would have it, an attempt to identify the wearer of any particular weave or colour by clan.

DAILY TONIC

One of the great banes of colonial life was malaria. It is still ubiquitous throughout much of India, Asia and Africa – it was only in the early 1950s that the disease was stamped out in the UK and the USA – and the only known medication was quinine which, in turn, was only to be found in the bark of the cinchona tree of the western reaches of South America. Quinine was very bitter to the taste so the colonials of British India, who were obliged to take a daily dose, took to dissolving it in a fizzy soda-based drink which they called their daily tonic and, just to jolly up the occasion, they added a slug of gin to help the medicine go down. Voila, the gin and tonic was born, with quinine still present in most brands of tonic water. But, with demand outstripping the bark from the aforementioned trees, the race was on to discover a synthetic form of the drug – which is where William Henry Perkin (1838–1907) comes in. The son of a London carpenter, wise in advance of his years, Perkin was just fifteen when he took employ as an assistant to Professor August von Hofmann (1818–92) in London's Royal College of Chemistry, now part of Imperial College. Part of his duties was to help his boss in his quest for a synthetic quinine. As it happened, this would not be successfully achieved until 1944 and then by Americans Robert B. Woodward and W. E. Doering of Harvard, but this did not stop Perkin trying to overtake his boss in the race.

In the Easter break of 1856, Hofmann said he had had enough and was going home to Germany for a while, leaving Perkin to his own devices. Working round the clock in a makeshift lab at his lodgings, Perkin tried time and time again, finally deciding to substitute simple aniline for the more complex potassium dichromate that had each time yielded nothing but a ruddy-brown sludge. The aniline yielded what was at first glance an even less appealing result – a thick black deposit. Deciding to give up himself and go to bed, Perkin went to wash out the flask with alcohol and was staggered to see everything turn a vivid purple – the sink, his hands – everything it came into contact with. He had accidentally discovered the first of the aniline dyes; he was soon a very wealthy man whom Hofmann never forgave.

THE BRITISH MONOPOLY

The next lucky accident in the synthetic dye industry, as started by Perkin, happened in the laboratories of BASF when Eugen Sapper (1858–1912) dropped his thermometer, which not only gave BASF the heads-up on how to synthesize indigo dyes in commercial quantities, it also sounded the death knell for the so-called British Raj in India.

The British stranglehold on their Jewel in the Crown extended to the indigo dye crop, which, by the late nineteenth century,

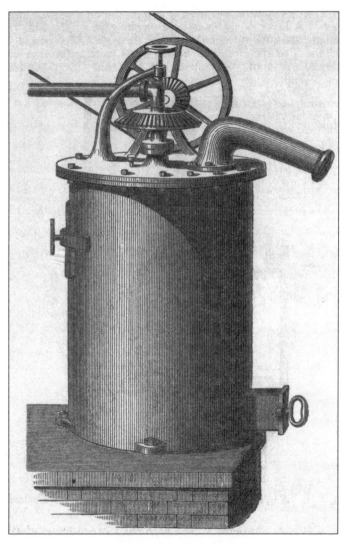

Apparatus for making aniline

they had bullied native planters to expand to an incredible three million acres. Crops such as tobacco or rice were 'discouraged' in vast reaches of Bengal by such brutal methods and tactics that it was rightly said that 'not a chest of indigo reached England without being stained with human blood'. An incredibly popular dye, the British made millions from the business while paying the growers next to nothing. Adolf von Baeyer (1835–1917) of the University of Berlin had been prompted to research the structure of indigo to see how it could best be synthesized, and so break the virtual monopoly held by the British. This he achieved by 1883 but none of the synthetic indigos could be produced within a cost-structure to rival the natural product. This work would nevertheless make him, in 1905, the first Jew to be awarded a Nobel Prize.

Digging lands for indigo

BROKEN GLASS

Next came BASF's Karl Heumann (1850–94) who was trying to find something useful to do with all the coal-tar generated by the steel industry; he extracted naphthalene from the foul-smelling gunk and used that as an organic base for his synthesis which, although more generous than anything that von Baeyer managed to generate, was still not a high enough yield to challenge the British on price. The next player in the game was the lab-rat Eugen Sapper. Little is known about him other than it was he who broke the thermometer that kicked the British out of India. Monitoring a batch of Heumann's naphthalene that was bubbling away with sulphuric acid in 1897, Sapper thought he would take the temperature but, instead, broke the glass tube against the side of the steel container. The mercury allowed the naphthalene to convert to phthalic anhydride, while the mercury itself was morphed into mercury sulphate by the sulphuric acid, which facilitated the production of indigo in vast quantities.

All of a sudden those millions of acres of indigo were worth nothing. But the British kept demanding the tithes and taxes from the very people from whom they were now refusing to buy the crop. The crop that they, the British, had forced them to plant in the first place. The worm turned; there had been Indigo Riots before but this time the plight of the indigo farmers became a rallying point for diverse groups across India, with

Gandhi taking up that cause in 1917 when he urged everyone to follow his lead and wear nothing but un-dyed cloth. Thirty years later, the little chap in his bed-sheet showed the British the door.

THE EXPLODING CAKE

Most assumed that the salts of picric acid would be explosive but no one thought the acid itself would be any use in that direction until 1873 when a German chemist called Hermann Sprengel (1834–1906) was running some through his self-designed vacuum pump. Actually, the Sprengel Pump was so efficient at creating vacuums that it could reduce the air in any chamber to one-millionth of its original status, this making possible the manufacture of filament light bulbs by Edison. The acid became a great favourite with the munitions trade and validation of the assumed high-explosive nature of picric salts came by accident in 1916. At La Pallice industrial complex at La Rochelle in France a vat of picric acid ruptured allowing the contents to spread across the floor, reducing and concentrating in the process. When the resulting shock-sensitive 'cake' was given an exploratory kick by an unfortunate worker, the ensuing explosion carried off 170 of his workmates.

EX-PAT'S 'EX-LAX'

One of the many synthetic dyes to emerge from the late-nineteenth-century German obsession with the industry was phenolphthalein, a compound with the ability to turn other fluids a rather fetching purple-red. In the opening years of the twentieth century, a vicious pestilence ravaged the Hungarian wine-grape harvest, presenting the producers with little option but to import other varieties to make up the shortfall and try to meet just the domestic demand. But these imported and lighter varieties failed to produce the blood-red results for which Hungarian wines are famed so phenolphthalein was added. Until then, no one had even contemplated using it as a food additive so no one was prepared for the results – which left almost the entire Hungarian nation with a most fearsome bout of diarrhoea.

Hungarian ex-pat, Max Kiss (d. 1967), read about this in his Brooklyn home and, blending the dye with chocolate, came up with a new over-the-counter laxative he originally called Bo-Bos. Shortly after this he was again reading a newspaper at home which this time carried news of an intransigent debate in the Orszaghaz – the Hungarian Parliament – and Kiss noted the use of the phrase 'ex-lax', the Hungarian slang shorthand for a political impasse. It was perfect; a blockage, albeit political, and an expression suggestive of 'excellent laxative', so he changed the name. The product is still on the market but that synthetic phenolphthalein has long since been replaced by natural senna.

HELLO YELLOW

The synthesizing of yellow dyes presents its own unique problems – with some attempts at their production nearly blowing their would-be creators sky high. Picric acid was first synthesized from natural indigo by the Irish Peter Woulfe (1727–1803) who used nitric acid in the extraction process. Not only was it a highly effective yellow dye but it also had antiseptic properties – the dyestuffs industry would effectively become the forebear of the pharmaceutical industry.

CANARY GIRLS

Germany's Joseph Wilbrand's (1839–1906) foray into the wonderful world of dyestuffs was also explosive. He was also hunting for a nice yellow dye when he came up with something called trinitrotoluene in 1863; it was a pretty enough colour but there were serious fixing and fading problems that resulted in it being shelved. About twenty years later, he was giving his laboratory storerooms and cupboards a thorough spring clean when some old containers of this yellow dye were chucked on the bonfire, with ear-splitting results – trinitrotoluene being better known today as TNT. This found great favour with the military of the day as it was extremely stable and needed temperatures in excess of 240°C to set it off – in other words it took a blasting cap. This meant it could be boiled up like soup in factories and poured into shells by munitions workers, who soon succumbed to its originally intended purpose. Known as Canary Girls throughout the First World War, these munitions workers not only turned a fetching shade of yellow but any with ginger hair had to endure it turning green. Some of their more adventurous workmates began using TNT to dye their hair a sort of blonde. (Hair that is dyed blonde today will also turn green if brought into contact with chlorine in a swimming pool.) Long-term exposure, on the other hand, was no laughing matter with the blood, spleen and liver function all suffering detrimental effects, as indeed did the immune system in general.

Mustard Gas

Last used by Saddam Hussein in the 1980s against the Kurds – and possibly by the Sudanese government against insurgents of the late 1990s – so-called mustard gas may be a bit old-hat but it is still hideously effective in its designed purpose. Actually, it is neither mustard nor gas but liquid chloroethyl-sulphide that is dispersed in atomized form, like a perfume-spray, by artillery shells, missiles or good old-fashioned aerial bombs. But the first use of mustard gas in the First World War is by far the most notorious.

Mustard gas was accidentally discovered way back in 1882 by César-Mansuète Despretz (1798–1863), who was tinkering with the interactions of sulphur dichloride and ethylene in a philanthropic effort to discover a way to rid the Third World of locusts. Luckily he decided the results to be too toxic and corrosive to deploy otherwise he might have rid the Third World of the Third World itself. But others picked up on his work, including the British Frederick Guthrie (1833–1886) who messed around with the compound in 1860, with Viktor Meyer (1848–1897) publishing a paper in 1886 after combining chloroethanol with an aqueous solution of potassium sulphide before jazzing up the mix with some phosphorus trichloride, just for good measure. Meyer had magnanimously tested the results on one of his assistants who ran around screaming in such an unmanly manner that Meyer thought he was hamming it up. But, after killing a few rabbits with the mix, he was convinced.

A MIX UP

The final tweak came in 1913 when Hans Thacher Clarke, a Brit working in Berlin with Emil Fischer, decided to replace the trichloride with hydrochloric acid and set the mix aside while he was distracted on other projects. Sometime later, as Clarke lifted the now matured mix from its resting place, intending to dispose of it, someone shouted to him from across the lab causing him to drop the flask, nearly killing himself and everyone else in the room. As Clarke was rushed to hospital, where he would remain for nearly three months, Fischer, who had witnessed the accident, was on the phone to contacts in the German Army, which had a production line in full swing before Clarke was back on his feet.

A poster published by the Chemical Warfare Service in 1915

It was at Ypres where the German Army first deployed mustard gas in October 1917, and both sides continued its use until the end of the conflict, by which time doctors were beginning to notice certain side effects in the gassed troops brought back from the Front. Apart from anything else it inflicted on those it encountered, the gas was also highly effective in suppressing the production of blood in the body.

SECRETS AND BOMBSHELLS

The war came to an end but the assimilation of case notes continued and, by 1942, with the Second World War in full swing, Yale University was already experimenting with the less-toxic nitrogen mustard and finding it highly effective in arresting the progress of lymphomas. But the two scientists running the programme, Louis S. Goodman and Alfred Gilman, couldn't tell a soul as their programme was under the aegis of the US Army and thus subject to complete secrecy.

But then the people who had started all the trouble in the first place came to the rescue, albeit unintentionally. Determined to slow the advance of the British 8th Army, Field Marshal Wolfram von Richthofen (1895–1945), cousin of the Red Baron and already notorious for his bombing of Guernica during the Spanish Civil War, decided to hit the

Allied fleet as it lay at anchor in the Italian port of Bari. On 2 December 1943, over 100 Junker 88s mounted the raid that the Allies would call Little Pearl Harbor and destroyed most of the port and every ship therein. But in the pall of choking smoke there was the smell of an old friend; some detected it as mustard while others said garlic. Survivors were, unfortunately for them, left in their clothing and just given blankets as they huddled in hospital corridors or holding areas, so it was not long before the doctors started to see the first symptoms: blistering edemas, skin falling off in great hanks and those proclaiming themselves to be feeling fine just dropping dead. The Germans, it seemed, had been dropping gas bombs.

STRANGE SMELLS

As the death toll rocketed in both Service survivors and the locals, doctors were at a loss and, eventually, on 7 December, Eisenhower detached chemical warfare expert Dr Lt Col. Stewart F. Alexander (1914–91) from his staff in Algeria and sent him to appraise the situation. One of the first things Alexander noticed on arrival at Bari was the garlicky smell that nervous officials suggested might be from the Germans having hit a garlic storage warehouse. But, given the fact that the smell was stronger inside the hospital wards than it was outside, Alexander was quickly on to the fact that mustard gas was the cause.

Because no one would talk to him or tell him anything pertinent, Alexander decided to use the epidemiology mapping methods first drawn up by the British Dr John Snow. Alexander obtained harbour plans and plotted the positions of all the ships, with pins used to represent the locations of those fished out of the drink after the attack and those on *terra firma*. Next he factored in the nature of the blistering on each patient, its pattern and location on the body. Then, by finally augmenting the data with the severity levels of the blistering on each person, he pointed the finger at the Liberty Ship *John Harvey* as the source of all the trouble. Sure enough, the British Staff officers in charge of the port finally admitted to Alexander that the *John Harvey* was full of mustard-gas shells when it blew sky-high in a giant fireball. Alexander was further shocked to learn that the load was there at the explicit instructions of his own boss, Eisenhower, who wanted it kept on hand to fire at the Germans if they so much as hinted at its deployment.

A CURE FOR CANCER?

Later there was anecdotal evidence of locals with varying cancers going into remission, but as the civilian population had largely and quite sensibly fled to other towns, this was difficult to evaluate. But there was no shortage of bodies for

Alexander to work with and all autopsies showed profound lymphopenia, a serious drop in the white blood-cell count, as well as suppression of myeloid cell lines. In the case of lymphomas and leukemias, there is a fatal overproduction of white blood cells by the diseased bone marrow so, in the light of the Bari autopsies, why not, Alexander reasoned, use the damned stuff to reduce such aberrant activity to normal levels?

It would also be later established that mustard gas dramatically inhibited or eliminated the kind of rapid cell division undertaken by growing tumours. No one would know much about DNA for years to come, but mustard gas prevents DNA uncoiling in the first step to cell division. All this information was fed back to Goodman and Gilman at Yale, who were left with the basic problem of finding a way of using this highly aggressive cytotoxic substance to kill a cancer before it killed the host.

BREAKING THE SILENCE

Fortunately for us, serendipity was not yet done with the team, when they decided to move onto animal trials. They selected a mouse at random and transplanted a lymphoma of such a size that, if left untreated and under normal circumstances, the host would have been expected to die within two weeks. Injecting a nitrogen mustard solution, the team saw the tumour shrink daily and the mouse rally to live another three months. Excited, they expanded trials, although they never replicated that first success. They would later admit that had they selected a different first subject who had died on them in the timeframe of all the subsequent mice, then they might well have abandoned the programme. But they persevered, moving on to human patients in trials that, like the animal ones, varied dramatically in the results. As Goodman and Gilman refined the agent and recalibrated the dosage, results improved to such an extent that the Military was forced to lift the pall of secrecy on the programme and allow, in 1949, Mustargen to become the first anti-cancer agent approved by the American FDA and like agencies across Europe.

Even now, nearly seventy years later, the echoes of the Bari raid reverberate still with mustard gas-based agents a standard treatment in many forms of the disease.

THE FEMININE TOUCH

In 1914, the American company of Kimberly-Clark found a way of producing highly absorbent cotton-like fibres from wood pulp and won contracts to supply the Allied Forces with disposable gas-mask filters and individually wrapped and sterile field dressings. The war suddenly over, Kimberly-Clark was left sitting on thousands of tons of such products until some bright spark decided to check through the bundles of letters sent in from nurses at the Front, who had wanted to thank the company for the product's roles in 'alternative' use. Apparently the gas-mask filters made admirable face-wipes, make-up removers and throwaway hankies, so they were re-packaged and marketed as Kleenex. As for the padded field dressings, the nurses wrote that they too came in handy for alternative use at certain times of the month. Prior to the First World War, they, like all other women, had to pack their undies with any old piece of cloth or rag to hand, so the field dressings were hurriedly re-branded as Kotex, the first such product available.

Penicillin

No one knows who first discovered the curative powers of moulds; in all likelihood there was no one single discovery, rather the chance observation of injured people grabbing anything to hand to staunch their wounds and a broad and slow-growing realization that those who used mouldy bread seemed to do better than others. The Ancient Greeks, Serbians and Indians all record the deliberate use of mouldy bread but, again, all such records imply that they knew nothing of the mechanism nor opted for any particular mould. The first mention of people deliberately growing a specific mould on a particular host for medical use comes from Sri Lanka, where records show that soldiers in the army of King Dutugemunu (ruled 161–137 BC) would set aside patties of an oil-based cake when battle seemed imminent and then take the mouldy results on campaign to be used as field dressings.

P IS FOR PADDINGTON

We know from the writings of Henryk Sienkiewicz (1845–1916) that it was the accepted practice in early seventeenth-century Poland to harvest spiders' webs that were contaminated with spores and mash them up with damp bread to serve as dressings; he explains the practice and gives the date reference in his *With Fire and Sword* (1884); he also wrote *Quo Vadis* (1895), filmed in 1951 as a sword-and-sandal epic. There are countless seventeenth- and eighteenth-century references to moulds in general being used to treat infection but the first solid, pre-Fleming reference to penicillium crops up in the 1809 writings of the German scientist, Johann Link (1767–1851), who not only coins the name from the Latin for a painter's brush, reference to the frond-like structure of the mould, but also describes three specific species: *P. candium*, *P. expansum* and *P. glaucum*.

On the domestic front, Sir John Scott Burdon-Sanderson, Medical Officer for Paddington – a place-name that seems to run a thread through the story – wrote in 1871 that the presence of penicillium inhibits the growth of bacteria. In that same year, a nurse at King's College Hospital came down with a hospital-acquired infection that was non-responsive to antiseptics so she was given alternative treatment and, when recovered and discussing her case, Joseph Lister's Registrar informed her that she had been treated with something called penicillium.

Across the Channel, a young French Army Medical Officer, Ernest Duchesne (1874–1912) was in 1897 presenting for his doctorate a thesis entitled *On the Antagonism between Moulds and Microbes* which had been inspired by his chance observation of the army's Arab stable-boys always choosing to store their saddles in the darkest and dampest place they could find to deliberately grow mould on the padded and upholstered underneath. When questioned by Duchesne they said it was an ancient practice where they came from and that the mould would automatically cure any sores that chaffed up on the horse. Duchesne took some of the mould and, having infected a batch of rats with various diseases, including typhoid, injected selected ones at random with *Penicillium glaucum* and, in all cases, these were the only survivors of the trials. Because he was so young and totally unknown, the Pasteur Institute did not even bother to acknowledge receipt of his paper – and all of this while, back in the UK, Alexander Fleming (1881–1955) was still running round in short trousers.

WHEN LIFE GIVES YOU LEMONS

Fleming had spend the First World War at medical stations in Northern France where he became increasingly alarmed at the death rates from infection alone – about half of the 10 million fatalities died not from bullets or gas but from assorted infections and disease – and this prompted Fleming to embark on something of a personal crusade against bacteria in general. This resolve was further strengthened by the ensuing 1918 flu epidemic that racked up double the death toll of the war in a mere six months.

Chance first nudged Fleming in the right direction in 1923 when it gave him a terrible cold while he was examining various cultures in the St Mary's Hospital laboratories in Paddington. Bent over his bench, his nose dripped into one of the Petri dishes prompting him to quit the lab, sneezing and cursing. On his return he was astonished; the location of the nose-drip was clear of all culture. Just to make sure, Fleming clamped one nostril shut and blasted the other at another dish which, in next to no time, was also cleared of all bacteria, albeit in something of a splatter-pattern. He summoned his assistant, the unfortunately named V.D. Allison, and repeated the procedure with Allison recalling: 'To our surprise the opaque suspension became in the space of less than two minutes as clear as water.' Fleming had discovered lysozyme, a naturally occurring antibacterial found in tears, saliva and nasal mucus.

Experiments continued, tears having been voted the more acceptable source, with V.D. Allison roped in for sore-eye duty: 'For the next five weeks my tears and his were our main supply of material for experiments. Many were the lemons we had to buy to produce all those tears. We used to cut small pieces of lemon peel and squeeze them into our eyes … and collect the tears with a pipette.' (Doubtless these men were awfully clever but why not just peel an onion?) Luck, like lightning, rarely strikes in the same place twice, but the gods of chance were not yet finished with Paddington.

EXPLORING OTHER CULTURES

The next step in the story involves a concatenation of extraneous factors; Fleming's notorious untidiness and laxity of method, his schoolboy sense of humour, the fact that he was something of a frustrated artist and a shot-in-a-million contamination to one of many samples he had forgotten to put through the sterilization process before going on holiday.

Different strains of staphylococcus mature to present in different colours and Fleming liked to make little designs in his Petri dishes – apparently he once managed to produce a Union Jack – and in the July of 1928 he walked out to go on holiday, leaving his bench strewn with the paraphernalia of research, which included about forty Petri dishes still with their culture samples. When he returned on 3 September, he gathered up these dishes for sterilization and dumped them, higgledy-piggledy, in a bucket to be doused in disinfectant. On impulse, he picked one out at random, to see if any interesting design had presented itself, and instead noted a blob of mould around which all the bacteria were dead and gone. The mould turned out to be quite rare, *Penicillium notatum*, and was traced to the lab on the floor below where they were experimenting with moulds from the homes of asthma patients in an attempt to desensitize them. A single spore must have escaped, hitched a ride on a stairwell draft and finally wandered into Fleming's lab and settled on that one single dish he had picked out of the jumble at random.

GOLD IN THE MOULD

And one would be forgiven for thinking that that was that; Fleming the bloodhound on the trail of the grail he had been chasing since the First World War – but no. Contrary to popular perception, the light never switched on in Fleming's attic. He found serendipity's gift interesting, but nothing more. He tinkered with the mould for a while but found it difficult to produce, impossible to stabilize, and finally dismissed it as useless in the fight against infection in humans. Fleming turned his mind to other matters but, about ten years later in 1939, Ernst Boris Chain (1906–79), a Jewish biochemist, fled Nazi

Germany for Oxford's cloistered serenity where, rooting through some old papers in storage, he came across a reference to Fleming and his unenthusiastic approach to penicillia in general. However, something stuck in Chain's mind and, when he was teamed up with the Australian Howard Florey (1898–1968) of the Oxford University Pathology Department, the two men re-trod the Fleming pathway. They failed to agree with his findings, but they both felt there was gold in that mould.

To be fair to Fleming, Oxford had far more sophisticated facilities than the labs at Paddington, so within two years Florey and Chain were able to isolate, concentrate and purify the mould to such a degree that they were ready for animal trials. And again chance intervened. They put in a request for a couple of guinea pigs, which, as the department was fresh out of guinea pigs, was fulfilled instead with mice. Undeterred, they went ahead and injected the mice with lethal doses of streptococci and then successfully treated them with their first sample batch of penicillin. What they could not have then known is that penicillin is fatally toxic to guinea pigs which, had they been delivered, would have died of the cure, leaving Chain and Florey doubting the validity of penicillin after all. Best of all, the urine of the two mice was permeated with the excess penicillin from the injections – mice are very small to take an injection intended for a guinea pig – proving that the drug could pass throughout the body fighting infection wherever it lay.

THE FIRST PATIENTS

There were further animal trials on a larger scale and, by January 1941, Chain and Florey felt ready for human trials. The first was an unnamed woman with terminal cancer who agreed to test the drug for its toxicity to humans. Although she had a bad reaction this was proved due to the impurities in the sample and not the drug itself. Thus assured of the lack of toxicity, the next trial was on a local policeman called Albert Alexander (1897–1941) who had gone from a simple scratch to his face while pruning roses to full-blown septicemia. On 12 February 1941, Florey administered 200 units intravenously and Alexander showed an almost immediate improvement – but they did not have enough penicillin to maintain treatment and he died three days later. Although disheartened by his death the team at least now knew they had made a breakthrough; the only problem remaining was that of large-scale production.

OVER THE POND

The team was running a Heath-Robinson-style production line, growing the stuff in old bed pans and, with little or no interest from the domestic pharmaceutical industry, Florey decided it was time to go to America. Taking the risk of his samples falling into German hands, he flew to

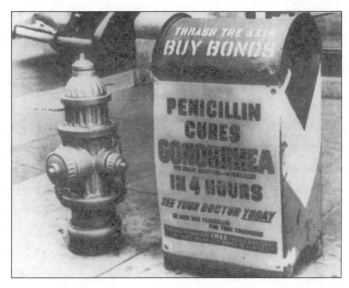

'Penicillin cures gonorrhea': advice offered to World War II servicemen

Lisbon – which although neutral was riddled with spies and agents – and thence to New York where a more enthusiastic welcome directed him to the Department of Agriculture in Peoria, Illinois.

Here he was introduced to Robert Coghill (1901–97) and, more importantly, Andrew J. Moyer (1899–1959), a microbiologist who is an all-too-forgotten hero of the story. Between them, Coghill and Moyer ran the Fermentation Division, which was trying to find a use for the millions of gallons of thick goop that came as a by-product of cornstarch production. Florey set to work and found that this incredibly rich base rendered

900 units per millilitre, whereas the best medium they had found back at Oxford only gave 2 units per millilitre.

THE LUCKY MELON

But the greatest lucky break was yet to come. Mary Hunt, a lab assistant under Moyer, came into work one day with a cantaloupe melon she had bought at the local market, only because it was contaminated with a rather fetching golden mould she had never seen before. And she had had the devil's own job of buying it from the vendor who, loath to have people think he sold rotten fruit, was all for chucking it in the bin and giving her a fresh one – free! But Mouldy Mary, as she is now known in the annals of medical lore, stuck to her guns, grappled the treasure from the hands of the tormented vendor and took it straight to Moyer.

Cultivating this new golden strain, Moyer further tinkered with the corn-syrup base, adding milk sugar, and thus increased the yield twenty-fold. Only through Moyer's input were the Allies able to have ready nearly 3 million doses of penicillin in time for the D-Day landings, this reducing deaths and amputations by over fifteen per cent.

When Chain and Florey first produced the stuff at Oxford the price per unit was beyond calculation; Moyer got that down to $20 per unit and, with his refined processes, by 1946 the price was a mere 50 cents. In 1945, Fleming, Chain and Florey shared the Nobel Prize for Physiology and Medicine, with Moyer not even getting a mention in the presentation speech. Many questioned the presence of Fleming himself, he being considered the least deserving of the three.

THE UNLUCKY DOCTOR

There is perhaps one last unsung hero in the saga: James Twomey, a doctor from Kanturk in County Cork, Ireland who, in the 1920s, was a partner at the Attercliffe Practice in Sheffield's industrial heartland. Recent research has uncovered that he routinely grew and administered his own penicillin with several notable successes. Whether this was a folk remedy he brought from his native Ireland or the result of his own endeavours we shall never know, for in 1938, and perhaps by then alerted to the work abandoned by Fleming, he visited London only to collapse in the street within a stone's throw of St Mary's Hospital in Paddington where Fleming was still at his bench. Barely conscious and unable to speak, Twomey died there on 17 May, in the same building that would have housed his cure, had he only been able to ask for it.

Catseyes

MOST NOCTURNAL or deep-sea hunters are blessed with a tapetum – a highly reflective layer, just behind the retina, whose job it is to fire the light back through the optic rods to afford its owner a double-take on every inbound image and thus enhance night vision. In the case of the cat, this has saved thousands of lives.

DRUNK DRIVING

British small-time businessman and inveterate tinkerer Percy Shaw (1890–1976) used the tramlines as guides when driving home from the pub: just keep the headlights on the shiny steel and you can't go wrong. But the trams stopped running and they dug up all the rails, leaving the more sober members of the community at the mercy of Percy and his like. One night in particular, in October 1933, Percy had been enjoying the odd Yorkshire ale in the Old Dolphin Hotel in Clayton Heights before driving home to Halifax in dense fog without the benefit of the rails to guide him. At one point, still believing himself to be on a straight section of road, the headlights picked up the eyes of a cat sitting on a wall, and the reflecting tapetum alerted Percy to imminent danger. He stopped and got out to realize that not only had he been on the wrong side of the road but that behind the wall was a sheer drop that would surely have been his end – Clayton Heights being about 1,000 feet above sea level.

Percy, now suddenly sober, continued his drive in deep thought and, by the time he arrived home he had a new bee in his bonnet: something had to replace the tramlines to make it safer to drive while drunk at night. He nipped out the following evening to 'liberate' some reflective studs currently used in local signage but soon realized that something altogether more sturdy was required if the cats' eyes assemblies were to withstand the pounding of traffic. And how best to keep them clean and operating at maximum reflectivity? He soon solved all these minor snags with the now-familiar design of a cast iron base and tough rubber shroud which, depressed by passing traffic, cleaned the face of the reflective lens with the wipers incorporated therein.

CAST-IRON CONCEPT

His first installation was at his own expense and involved his making safe the notorious accident black-spot of Drighlington Crossroads in Bradford by installing over fifty of his new products in April 1934. (There is one reference on the internet to a Jean Neuhaus installing his Follsain Gloworms for Market Harborough Urban District Council in March that same year and, if true, then Shaw was pipped to the post.) By 1937 the Ministry of Transport had woken up to the advantages of reflective road studs and conducted tests on several designs but, with the notable exception of Shaw's Catseyes, within a very short time all had failed, cracked or been dug out of the road by passing traffic. All those who tried to copy him failed on quality. Shaw, a true Yorkshireman, put his faith in cast iron and the best quality rubber available; he did not and never would compromise quality for the sake of a quick profit.

Shaw made a fortune, he could have made more by outsourcing manufacture to overseas providers but he did not trust foreigners who, in his opinion, began at the Yorkshire county boundary. During his lifetime all production rested in Yorkshire, where his private life was strange to say the least.

Percy Shaw supervising production in his factory
in Boothtown, Halifax in 1958

Living in a sizeable house, Shaw refused to have curtains as
he felt them to be nothing but dust-collectors that obscured
his view of the local countryside. Likewise he would tolerate
neither carpets nor rugs and kept four televisions turned on
in the lounge at all times. One tuned to BBC1, one to BBC2,
one to ITV and the fourth in case any of the others broke
down. If anything attracted his attention he would turn up
the volume for a while and then return the set to mute with its
fellows. He liked the ladies but never found anyone he trusted
enough to marry.

The Microwave Oven

IN THE SUMMER of 1945, Percy LeBaron Spenser (1894–1970), a senior technical manager at Raytheon, a military/navy contractor of Waltham, Massachusetts, was helping to prepare a demonstration of some new radar equipment to visiting military brass. With time short and no chance to get to the canteen, Percy grabbed a chocolate Mr Goodbar, but, fortunately for us, he was to be denied even that extravagance by the arrival of the visitors, which required Percy to hurriedly shove the guilty booty into a pocket in his lab whites. As the magnetron driving the equipment was fired up, Percy became aware of an embarrassing stain spreading across his trousers and he backed sheepishly out of the room to discover that the chocolate bar had not just melted but had actually begun to cook; it was still warm. With the departed guests suitably impressed by the equipment in its intended use, Percy returned to the lab with some popcorn which, placed in front of that same magnetron, was soon zinging all over the room. Moving on to what he felt would be a less dramatic experiment, Percy aimed the device at an egg and sat back. He recalls a curious colleague who wandered up and asked what he was doing, to which Percy replied that he was trying to boil an egg without water. Intrigued, the colleague moved in for a closer look just as the egg decided to explode, leaving them both with egg on their faces.

ALL THE RADARANGE

The company took out a patent on 8 October 1945 and branded them, rather ominously, Radaranges. But they were about the size of a commercial fridge, they required plumbing in to keep cool some fairly serious internal workings and, more importantly, cost a little over $3,000 – about $50,000 at today's values. Charles Francis Adams IV (1910–1999), Raytheon's CEO, had one installed at his expansive residence but this was destined to stand unused as his cook walked out saying she would have nothing to do with the damned thing. (Adams was also a descendant of America's original 'Adams Family', which included presidents John Adams and John Quincy Adams.)

And it must be said that market penetration in America was very slow, even though smaller versions with smaller prices emerged all the time. In the twenty years following Percy's brainchild having its birth registered in the Patent Office, a scant 11,000 were sold throughout the USA. Trouble was, Americans wanted seared steaks, burgers and chips, and other things that microwave ovens are still no good at conjuring forth.

But the Japanese diet *was* ideally suited to microwave technology and, in 1961, the Japanese National Railway led the way by transferring all station and rail catering to microwave ovens made and installed by Toshiba. Even street vendors were using them to re-heat previously prepared snacks and the domestic market absorbed about another 300,000 units per year. By the 1970s, Japanese domestic sales had rocketed to 1.5 million per year with the American market struggling to make the 250,000 mark.

Darwin: The Accidental Tourist

OF ALL THE ACCIDENTAL and serendipitous discoveries, innovations and advances in thinking, that which led Charles Darwin (1809–82) to eventually arrive at the conclusions he did has to be one of the most famous. But certain things should be understood at the outset: theories of evolution were nothing new in Darwin's day; musings on the subject date back to Ancient Greece, and even his own grandfather, Erasmus Darwin (1731–1802), among many other notable names, had written extensively on the subject. As for Darwin, he spent the entire voyage on HMS *Beagle* with his mind on other matters and, far from his having had the eureka moment of popular imagination on the Galapagos Islands, he spent his short and unenjoyable visit blithely ignoring what was staring him in the face and killing and eating the unique specimens that could have taught him so much – but then he was very young.

It is a fair bet that most readers, if asked to shut their eyes and conjure up an image of Charles Darwin, will have in their imagination a picture of a serious old chap with a long and bushy grey beard, glowering in a chair, or some such. But Darwin was still a rather spoilt and somewhat closeted twenty-two-year-old when HMS *Beagle* sailed and, far from anticipating a scientific expedition, he was at the time a devout Creationist aiming to settle down to the quiet life of a country pastor. He was not invited along as the expedition's naturalist but as a dinner companion for the Captain. He hated the Galapagos Islands where he largely ignored the unique flora and fauna in his search for geological samples. (All the myths built round the so-called Darwin finches and transmutation of species would come later.) And, apart from all that, chance had to work a great deal of its magic just to get him on to the decks of HMS *Beagle* in the first place.

FIRST SIGNS OF MADNESS

The story starts with the less-than-balanced Robert Stewart, Viscount Castlereagh (1769–1822), a close relative of Captain Robert Fitzroy (1805–65) who would later invite Darwin to join him aboard HMS *Beagle*. Ever-volatile, Castlereagh, over imputations of Castlereagh's mismanaging certain aspects of the Peninsular War, challenged the Foreign Secretary George Canning (1770–1827) to a duel, resulting in the pair facing off on Putney Heath on 21 September 1809. Although Castlereagh survived the encounter, he was roundly pilloried for having called out Canning in the first place and, with his reputation in tatters, he sank into an increasingly dark depression that only ended when he cut his own throat. This was the first intimation of madness in the family at a time when such condition was widely believed to hereditary, and Castlereagh's suicide was still very raw in the mind of the young Fitzroy when he took command of HMS *Beagle* six years later.

HMS *Beagle*

A LONG VOYAGE

The next to rush rudely unannounced into the presence of the Almighty was Pringle Stokes (1793–1828) who put Darwin a step closer to the gangplank of HMS *Beagle* when, as Captain of the same, he shot himself in the head on 12 August 1828. The ship was on a hydrographic survey of the waters around Patagonia and Terra del Fuego, so command passed to the meteorological officer, Lt Robert Fitzroy (1805–65). Still only twenty-three years old, Fitzroy had some heavyweight connections throughout society in general and the Admiralty in particular; the Fitzroys had made it their business to marry well to extend their influence well beyond the dynasty of the aforementioned and unstable Castlereagh family and into some of the richest and most powerful families in the land. The summer of 1831 saw *Beagle* refitted and upgraded, her wealthy new commander sparing no expense. Thus she stood ready for her more famous Second Voyage which would be another but more far-ranging hydrographic survey that required Fitzroy to take her down the east coast of South America, up the west coast, across to the Galapagos Islands, thence to Tahiti and on to Australia and New Zealand before heading for home. But the Black Dog was already sitting at Fitzroy's side; not only was he worried about hereditary insanity in his own family but he had all too recently seen what the lonely command of a long voyage can do to someone's mind – and HMS *Beagle* was scheduled to be away for over two years (and, in fact, she was away for five).

CAPTAIN SEEKS MATE

So Fitzroy wrote to a close friend, Admiral Francis Beaufort (1774–1857), he of wind-strength fame, asking him to recommend a suitable travelling companion who, rather than 'some dammed collector of specimens' should be the sort of chap that he, Fitzroy, could chat to on his own level. Beaufort knew exactly what his friend wanted and approached several suitable candidates at Cambridge University but all had better things to do with their time – all except the somewhat unruly Charles Darwin, who, despite being in training to be an Anglican parson, was more interested in riding, hunting and fishing. When Darwin stalked the countryside everything was an endangered species. At the bottom of Beaufort's list, he was, on the other hand, well bred and known to be at a bit of a loose end at the family home in Shrewsbury and so received a written invitation to join the fun:

> Capt. F. wants a man (I understand) more as a companion than a mere collector & would not take any one however good a Naturalist who was not recommended to him likewise as a *gentleman* . . . there never was a finer chance for a man of zeal & spirit . . . Don't put on any modest doubts or fears about your disqualifications for I assure you I think you are the very man they are in search of . . .

But Darwin senior was not kindly disposed to the notion of his wayward son dodging his studies to embark on some protracted gad-about, and he put his foot down; it was hardly the sort of stable behaviour befitting one so shortly destined to take the cloth. And Charles himself was not that bothered; it certainly sounded like a jolly jaunt but so too did the life of a rural parson who, with a well-ordered parish, would have plenty of free time to devote to the slaughter of God's creatures. Besides, this Fitzroy fellow was asking for a £500 down payment against feed and keep, so without Darwin senior footing the bill there was no way he could take up the offer anyway. Instead, Charles took himself off to the estate of his uncle, Josiah Wedgwood II (1769–1843), he of cup-and-saucer fame, for the beginning of the partridge season.

Between shots, Darwin made reference to Fitzroy's invitation and, unlike Darwin senior, Wedgwood was of a mind that this would be the making of the lad, so the very next day he badgered his brother-in-law to the extent that the poor chap was left with no option but to relent and cough up the stake money for Charles to join the cruise.

However, while the Darwin household stood divided against itself on the issue, Fitzroy, miffed at no reply either way, had offered the opportunity to a friend, Harry Chester, and written to Darwin telling him to forget all about it; fortunately this letter was still in the post and crossed with

Darwin on his way to London to discuss the matter and, by the time he arrived in London, Harry Chester had decided that such a protracted spell at sea would play havoc with his social life and changed his mind.

A LAST RESORT

But there were still the weighty problems of Darwin's politics and the shape of his nose. Not only was Fitzroy a Tory and Darwin a devoted Whig but the latter was also possessed of a nose that, to the dilettante physiognomist Fitzroy, clearly betrayed a lack of determination and fortitude. But departure was scheduled for that very month end – September – and the only halfway acceptable candidate sat before him: there simply wasn't anyone else; it was Charles Darwin or risk suicide, so Darwin it was.

Unbeknown to Fitzroy the trip was destined to be delayed for a further three months by various snags and hindrances, the last of which being the crew's Christmas drunkenness. And it is a good job for later science that no one could have anticipated those delays for, as Darwin himself would later record, Fitzroy would have left him high and dry on the dockside for the shape of his nose alone if he'd thought for a second he'd have had the time to find a replacement. But, with a seriously hung-over crew, HMS *Beagle* put to sea on 27 December 1831 when, with Dr Robert McCormick (1800–90) as the expedition's official naturalist and Darwin but the travelling companion to the soon-to-become unstable Captain, she struck out for the Azores. As stated, the trip overran by three years and the two men would argue, quite violently, from time to time, with an increasingly volatile Fitzroy stamping out of the dining room only to make abject apologies the next day.

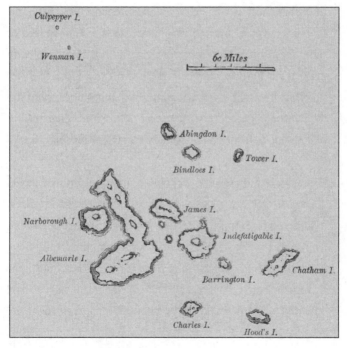

Galapagos archipelago

RACE RELATIONS

Doubtless fuelled by the amount of wine they had on board for the trip, one of Fitzroy and Darwin's fiercest rows was over the extensive slavery they witnessed in South America; Fitzroy thought it fitted the natural order of things while Darwin found it repugnant. But this should not be taken as any indication of Darwin as some warm and fluffy liberal, brimming over with the milk of human kindness; very much a product of his class and time, he would later muse in his *The Descent of Man, and Selection in Relation to Sex* (1871) that the down-side of medical advances was that they allowed the weaker and less admirable members of society to live long enough to breed more of their own repugnant kind. And, from the same book we get his opinion that:

> At some future period, not very distant as measured by centuries, the civilized races of man will almost certainly exterminate and replace throughout the world the savage races. At the same time the anthropomorphous apes . . . will no doubt be exterminated. The break will then be rendered wider, for it will intervene between man in a more civilized

state, as we may hope the Caucasian, and some ape as low as a baboon, instead of as at present between the negro or Australian (Aborigine) and the gorilla.

His diary on HMS *Beagle* notes the Maori to be 'filthy and offensive' and 'the scum of the Pacific'. It was this and other elitist musings that prompted his cousin, Francis Galton (1822–1911), to establish the Eugenics Movement, which would be taken to its illogical extreme under Hitler.

ROCKING THE BOAT

In due course, HMS *Beagle* rounded South America and, on 16 September 1835, arrived at the Galapagos Islands where, as most imagine and TV docu-drama would have us all believe, Darwin had an epiphany and forged the theory of Evolution, wholecloth, from evidence presented by the local fauna which, isolated from external influence and driven only by the dictates of its cloistered environment, had developed in its own unique forms. Well, nothing could be further from the truth.

Darwin, still a Creationist, was more interested in geological data and samples and pretty much ignored what was staring him in the face during his remarkably short visit to those islands. In a letter to his sister he eulogizes:

> . . . there is nothing like geology; the pleasure of the first days partridge shooting or first days hunting cannot be compared to finding a fine group of fossil bones, which tell their story of former times with almost a living tongue.

Fine praise for geology from a hunter such as Darwin but it is a matter of record that he spent a scant nineteen days ashore throughout the five short weeks that HMS *Beagle* was anchored there. In short Darwin learned nothing at all during this brief visit to the Galapagos, only later, and with the help of others who *had* labelled their specimens according to the islands on which they had been found, could all the bits of the puzzle fall into place.

TURNING TURTLE

On arrival at the islands, they were met by Acting Governor Nicolas Lawson who, had Darwin's mind been receptive to notions of evolution and natural selection at the time, gave him the greatest of all clues. Lawson's party-piece was to demonstrate how he could tell 'with certainty' on which island any of the famous Galapagos turtles had been born, simply by the shape of its shell, as the lifestyle on each individual island in the group produced variations, mainly in the projection of the carapace behind the head. Darwin, who would later note that such thinking 'undermined the [idea of the] stability of the species', ignored any mention of carapace variance from island to island and instead rounded up about thirty turtles for him and Fitzroy to eat on the next leg of their journey. The all-important shells were simply tossed overboard; far from rejoicing and noting their variation, Darwin ate the evidence. Nor did he have much to say of the Galapagos themselves other than that they were ugly, uninspiring and, with their black, jagged rocks infested with giant iguana and lava-caked interiors, 'what we might imagine the cultivated parts of the Infernal Regions to be'.

FOR THE BIRDS

Darwin did collect samples of the birds from the islands but, missing the point completely, failed to note from which island he took them – he also failed to identify them properly as his ornithological knowledge was mostly gleaned through a gun sight. And it was mockingbirds that first set him questioning his Creationist leanings, not the finches, but it would be months after leaving the Galapagos that he noticed the variance in the specimens and, still missing the point, he deemed them to be 'only varieties' rather than the true and separate species that they were.

The first thing Darwin did on his return was to present his beloved rock samples to the Geological Society of London at their meeting of 4 January 1837. His badly catalogued and identified avian samples were passed to ornithologist John Gould (1804–81) who, at the Society's next meeting on 10 January, reported that the birds Darwin had thought to be blackbirds, grossbills, wrens and finches were in fact *all* finches and, more than that, they were a series of finches so unique as to form 'an entirely new group containing twelve species'. Thus it was Gould who established the truth, not Darwin who, now realizing that island categorization was of prime importance, scurried round to beg, borrow or steal the island-specific samples collected by Fitzroy and a chap called Syms Covington (1816–61) who had been Darwin's servant throughout the trip. Gould also advised that the small rhea Darwin had so nearly eaten for his last Christmas dinner at sea, but three weeks before, was in fact another new species, thankfully saved from going the same way as the turtles.

FINCH BY FINCH

It would be a long time before Darwin realized the magnitude of what he had stumbled across with those famous finches not even getting a mention in his *On the Origin of Species* (1859) and only a passing reference in his *Journal* (1839). Only in later editions did Darwin make larger reference to the finches but, even then, the most he ever had to say about them was:

Seeing this gradation and diversity of structure in one small, intimately related group of birds, one might really fancy that from an original paucity of birds in this archipelago, one species had been taken and modified for different ends . . . Unfortunately most of the specimens of the finch tribe were mingled together; but I have strong reasons to suspect that some of the species of the sub-group Geospiza are confined to separate islands. If the different islands have their representatives of Geospiza, it may help to explain the singularly large number of the species of this sub-group in this one small archipelago, and as a probable consequence of their numbers, the perfectly graduated series in the size of their beaks.

So, Darwin was never sure that each of the Galapagos islands produced its own species of finch, the closest he got to that truth was his speculation that the different finches had all descended from a common ancestor and undergone minor changes to enable them to do different things, as dictated by their diet.

Fig. 2 Darwin's finches; the male (in dark plumage) and female of each species: *1, 2, 3*, the Large, Medium, and Small Ground Finches (*Geospiza magnirostris, G. fortis,* and *G. fuliginosa*); *4*, the Sharp-beaked Ground Finch (*G. nebulosa* [formerly *difficilis*]); *5* and *6*, the Cactus and Large Cactus Finches (*G. scandens* and *G. conirostris*); *7*, the Vegetarian Tree Finch (*Platyspiza crassirostris*); *8, 9*, and *10*, the Large, Medium, and Small Insectivorous Tree Finches (*Camarhynchus psittacula, C. pauper,* and *C. parvulus*); *11*, the Woodpecker Finch (*C. pallidus*); *12*, the Mangrove Finch (*C. heliobates*); *13*, the Warbler Finch (*Certhidea olivacea*); and *14*, the Cocos Island Finch (*Pinaroloxias inornata*). (From Lack, 1947 : 19.)

Darwin's finches

JOURNEY TO THE END

Far from his having a eureka moment on the Galapagos, it would be a long and slow process for Darwin's ideas to themselves mutate and evolve. He was supported through this endeavour by his wife, the quietly resilient Emma Wedgwood (1808–96), who was also his first cousin. Constantly worried about their close blood-ties, which went against all his darker ramblings on selective breeding of the human race, Darwin would go into fits of guilt whenever any of their children took sick, in case it was the first signs of inherent weakness occasioned by them being the product of such close inbreeding. As for Fitzroy, he followed in the family tradition by doing what he had hired Darwin to prevent during the voyage. On a bright and sunny Sunday morning, 30 April 1865, he quietly locked himself in his dressing room and cut his throat with a razor.

Pavlov's Dog

THE NAME OF IVAN PAVLOV (1849–1936) has to be among the best known from the realms of science, with most being familiar with the term 'Pavlovian response' and there being no fewer than three bands named after him and his lab-dogs: Pavlov's Dog was formed in 1972 in St Louis and Deep Six rebranding themselves as Pavlov's Salvation Army in 1982 and, lastly, the now-defunct bluegrass Pavlov's Dawgs beginning their ten-year run in 1988. Most remember him as having 'something to do with psychology' whereas he was in fact a physiologist exploring the digestive system of dogs. If he'd had his way he would have deftly side-stepped his 1904 Nobel Prize and been relegated to relative obscurity.

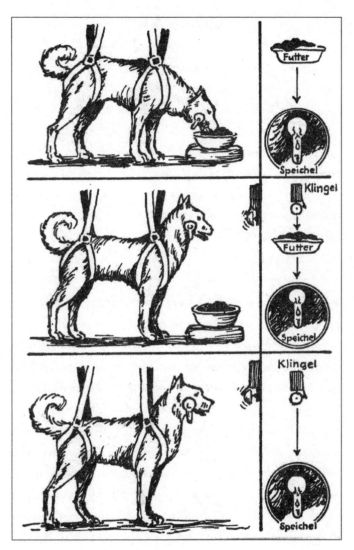

Pavlov's dog experiment

BUSTING A GUT

In 1890 Pavlov was invited to join the Institute of Experimental Medicine of the University of St Petersburg to run the Department of Physiology and by the end of that same decade he was well into his experiments on the gastric function of dogs. His unfortunate canine helpers had all been 'adapted' to the purpose with their saliva glands externalized through holes in their head to which were attached calibrated vials for the contents to be monitored throughout the day. Pavlov soon noted that the dogs began to produce saliva at the very sight of the food about to be fed them and, missing the point entirely, regarded this as highly disruptive to the programme and a problem to be overcome by a complicated series of ruses and mechanical devices to prevent the dogs from being aware of the impending treat.

But the dogs were not to be cheated. The feeding times were carefully set in order to time the passage of food through the gut, so alarm bells went off at regular intervals to remind the staff it was time for the next meal. It was the staff, not Pavlov, who noted that the dogs were now salivating at the time of the bell and told their leader, who promptly ordered the bell to be muted and, when even this did not prevent his charges from brandishing the chalice of international acclaim in his face, he had them all moved to soundproofed rooms. But the dogs were, well, dogged in their persistence and continued to salivate at the sight of a white coat or even at the footfall of the feeders that they had come to recognize. Then and only then did the dogs and staff win through, with Pavlov turning the slant of the programme to exploring what came to be known as conditioned response or reaction.

TEDDY SCARE

Inspired by Pavlov's success with dogs, the American psychologist John B. Watson decided to extend the programme to humans in a rather disturbing programme that came to be known, rather glibly, as Scaring the Crap out of Little Albert. In 1920, at the Harriet Lane Home for Invalid Children, part of Baltimore's Johns Hopkins Hospital, Watson and his assistant, Rosalie Rayner, picked on an eight-month-old child, known only as Albert B., and conditioned him to fear anything from a white rat to a glove or a teddy bear. Presenting the child with such items to establish there was no inherent fear, Watson would re-present Albert with the same items but, this time, strike a steel rod or a large gong with a hammer, right behind the poor little chap, until he cried his eyes out. Not surprisingly, Albert would cry in terror when confronted by a teddy bear, or whatever, even if the gong was not banged to scare the living daylights out of him.

Watson and Rayner (twenty years his junior and one of his students to boot) were also conducting trials on sexual response and orgasms using themselves as enthusiastic models but, when their casenotes came to light, Mrs Watson and the Johns Hopkins Institute failed to share their enthusiasm, each showing Watson their respective doors.

DOGS OF WAR

But the prize for the most bizarre use of Pavlov's findings must go to his compatriots in the Russian Army.

The standard tank used by the Russians in the Second World War was no match for the German Panzer, designed by Ferdinand Porsche (1875–1951), of later sports-car fame, who would be arrested after the conflict for his use of slave labour in his factories. Nor were Soviet anti-tank weapons much use against the superior armour of the Panzer so a new and terrifying weapon was invented. Thinking of Pavlov's results with dogs and food, the Russian tank regiments trained dogs to associate the underneath of tanks with food by starving them for a few days and then letting them out to see piles of meat and treats placed under stationary tanks with the engines running.

The idea was as Pavlovian as it was brutal; cement the tank-food association and then starve the dogs, strap mines to their backs and let them loose on advancing German armour. But the Pavlovian link was too firmly forged in the minds of the dogs that only associated the underneath of *Russian* tanks with food and, on their debut in 1942, packs of tooled-up dogs forced three brigades of Russian armour into a panicked retreat. It was later figured out that this was likely because the Russian tanks smelled of diesel whereas the German tanks ran on petrol but by then it was all too late – the Pavlovian Dogs of War were retired.

Post-it Notes

In 1968, Dr Spencer Silver (b.1941), Senior Chemist at Minnesota Mining and Manufacture's Central Research Facility came up with an intriguing but apparently useless new adhesive while trying to find a new page-binding process for the publishing trade. In all experimental batches the pages clung together – just – but came apart, undamaged, at the slightest tug. Ever the optimist, Silver remained convinced that there simply *had* to be a commercial application for a low-tack, reusable adhesive but no one else in 3M shared his enthusiasm.

SINGING FROM THE SAME HYMN SHEET

Scroll forward to 1974 and we find another luminary of 3M's research facility, Art Fry (b.1931), with problems arranging the musical sheets and prompts for the choir at his local church. If he glued the memos and prompts into the hymnbooks or on to the score sheets they had to stay in place forever lest the pages be torn with their extraction or, if inserted loose, they fell out and fluttered around the altar during recitals. Then he remembered Silver's abortive foray into the bookbinding market and was soon using what we now call Post-its for all choral events.

STICKING IT OUT

Back at work Fry then took up the gauntlet dropped by Silver and tried to raise some commercial interest but he too failed, with 3M marketing gurus hiding firmly behind the basic question of what need is fulfilled by these sticky-edged bits of paper? All very clever but who is going to buy these and what are they going to do with them? Fortunately for Fry and the rest of us, 3M have always allowed researchers a certain amount of what they call 'bootleg time' during which they can pursue any project they choose. Fry chose to use his bootleg time perfecting the product and, since *he* could not think of any particular use either, he finally managed to convince the

Office Supply wing of 3M that his sticky-pads were destined to 'meet an as-yet-unperceived need'; the guy should have been a politician.

Finally, in 1978, 3M Office Supply gave in and out went hundreds of the little yellow pads to every customer they had in Boise, Idaho, a town chosen by the time-honoured expedient of shutting the eyes and sticking a pin in the map.

Fry did not have to wait long for the feedback. It came in the form of frantic phone calls for fresh supplies and massive orders for his brainchild, which had been stuck to every surface imaginable in the office with scribbled notes from one colleague to another. Most encouraging of all was the fact that half the initial drop had been pinched by employees who put them to use around the home and even in the car. The Post-it note had arrived – but this was not 3M's first venture into strip-edge adhesives. In fact, it was a casual conversation in a garage that was overheard by chance that got them into the tape business in the first place.

MAKE DO AND MEND

Back in the 1920s, when abrasive paper was 3Ms staple product, the then leading researcher Richard Gurley Drew (1899–1980) was collecting his car from the body shop, after some minor prang, and he heard the paint-sprayers lamenting the increasing popularity of two-tone bodywork and the lack of any adequate masking tape to achieve a clean line between the two colours. It seems that all the tapes on the market were either so sticky that they tore away paint when removed while the less sticky ones were all backed by cloth, which absorbed solvents from the spray-job to leave an imprint of the weft and warp of the fabric. Pondering the problem, Drew came up with a broad tape of crepe paper which only had a line of low-tack adhesive down each edge to reduce problems in the spraying process and subsequent drying. Trouble was, Drew was a shade too parsimonious with the adhesive and the tape kept 'falling down' on the job, prompting the ungrateful sprayers to apply the nickname of 'scotch tape' for the stingy line of adhesive. The application of adhesive was expanded across the face of the tape whereupon it stuck as fast as the nickname which 3M adopted and registered. New transparent cellophane versions were soon marketed but it took the Wall Street Crash of 1929 to launch the product to the general public, who were then forced to mend things rather than discard them.

And an accident of a different kind at 3M is also responsible for one of the biggest bus companies in the world.

A NEW BUS-INESS

In 1914, the Swedish-born Carl Eric Wickman (1887–1954) was a driller in the 3M mines but, due to an accident at work, was invalided out of the company and, after an unsuccessful period of trying to sell seven-passenger people carriers for Hupmobile, he decided to buy one himself and ferry his old workmates to and from the mines at 15c a go. He soon had to buy another – and another – and in no time at all found himself running an ever-expanding fleet. Everything around the 3M mines was covered in grey dust and, because Wickman's Hupmobiles moved people around quickly, they acquired the nickname of greyhounds, with Wickman living to see his company grow into one of the most famous names in the bus business.

A Cure for Scurvy

ALTHOUGH ENGLISH NAVY surgeon, Dr James Lind (1716–94), is roundly credited with discovering the cure for scurvy in lime juice this does seem to be a trifle unfair, if only for the fact that he was at the time trying to prove that acid – any acid – was the cure; he was in fact completely ignorant of the medicinal benefits of fresh citrus fruit, or indeed of vitamin C in particular, which would not be discovered for decades. Not only that, but others had for centuries been advocating the use of vitamin C-rich plant life. Lind himself was dead before Western physicians became aware of the real cause of the disease, which the despised savages of The Americas and India had known for centuries. So, quite how Lind managed to grab all the kudos is a mystery in itself.

BARKING MAD

An extreme form of vitamin C deficiency, scurvy was first discussed by Hippocrates (460–370 BC), he of the oath that doctors no longer take or subscribe to. Hippocrates diligently noted the condition's manifold manifestations and deadly effects while remaining ignorant of the cause. The first recorded incidence of plant-matter rich in vitamin C being used to cure scurvy is probably the 1536 exploration of the St Lawrence Seaway by Jacques Cartier (1491–1557) who was on the point of losing his entire crew until the locals showed him how to brew up a preparation from the needles of the white cedar tree. A similar intervention saved Sir Francis Drake's (1540–96) crew during his circumnavigation when, in 1577, with his crew ravaged by scurvy, he put ashore at what is now Patagonia. Here again the natives saved the interlopers with a mash made from the bark of a local tree, also high in vitamin C. Unfortunately for thousands of sailors and soldiers to come, the medical minds of both France and England scoffed at the reports; what did Johnny Jungle know?

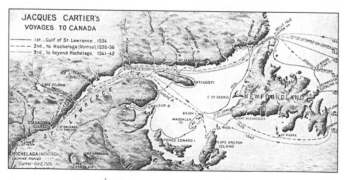

Cartier's exploration of the St Lawrence Seaway

RUM RATIONS

By the opening of the seventeenth century there was informed opposition to the conventional 'wisdom' of the day, which held that the condition was curable by the oral administration of dilute acids; John Woodall (1570–1643), Surgeon General of the East India Company who never dismissed out of hand the traditions of the lands in which he was posted, wrote in 1614 that it should be combated with fresh food or, if none was available, plenty of oranges, lemons or limes. In 1734, thirteen years before Lind conducted any of his misguided experiments, the Polish-born theologian-physician Jan Fryderyk Bachstrom published *Observations on Scurvy* in which he stated that: 'scurvy is solely owing to a total abstinence from fresh vegetable food and greens, this alone is the primary cause of the disease.'

Lind was aware of these opinions but, still a staunch member of the 'acid lobby', he was galvanized into serendipitous activity by anecdotal evidence from the Caribbean Fleet which seemed to enjoy a much lower than average incidence of scurvy since its Commander-in-Chief, tired of dealing with the consequences of the excessive drinking in the navy of the day, had ordered the daily rum-ration to be cut with lime juice. This was Admiral Edward Vernon (1684–1757), whose heroics at Porto Bello during the War of Jenkins' Ear inspired the writing of 'Rule Britannia'. Before Vernon's initiative of 1740, standard issue was 95%-proof rum, served at the knee-buckling rate of half-a-pint per man per day, and this on top of extra issues for work well done and beer being drunk routinely through the day – water could be dangerous when stored on board for long periods in such hot climes. Vernon ordered the ration to be reduced in quantity, split into two issues and diluted with pure lime juice. Not only did this render his men less drunk than any others afloat but the incidence of scurvy seemed to drop in tandem.

TRIAL AND ERROR

This new ration was not popular with the floating knees-up that was the eighteenth-century navy, the men nicknaming the dilute ration 'grog', this being Vernon's own soubriquet promoted by his heavy and distinctive boat cape made of a coarse material called grogram. But the most far-flung effect of this move to relative sobriety was its serving as the inspiration of Lind to prove that it was the acidity of the lime juice that was responsible for the reduction in scurvy throughout Vernon's fleet; his headlong dash to prove the fresh fruit namby-pambies wrong in fact asserted just the opposite – he also invented the clinical trial, by accident, while he was at it.

Gathering a dozen established scurvy cases to his ill-informed bosom, Lind divided them into six pairs, each pair to be treated differently. The first pair was given daily measures of cider, the second hot spices in barley water, the third pair vinegar, the fourth pair was forced to drink seawater, the fifth and lucky pair, Lind's favourites, got the fresh oranges and lemons and the really unlucky sixth pair endured a daily dose of dilute sulphuric acid; you can bet that hurt in the morning. After six days, the frugivorous pair was back at work, forcing Lind to close down the trials as he had simply run out of fruit and thought it unethical to continue with one comparison group eliminated. Besides, Lind was quietly convinced that the early return to work of that pair was validation enough of his fruit-acid theory.

ACID TRIP

Actually, while Lind had deliberately structured the trials in that manner to give his pet theory the best chance to shine through, he had nevertheless laid out the basics for all blind and double blind clinical trials to come. Not only that, but he had failed magnificently to grasp the chalice; still he adhered to the notion that acids alone would purge the body of this malady and the results of his trials only served to convince him that citric acid was the most efficient of acids. Still he failed to recognize citrus fruit as the prophylactic discussed by so many before him. And this false premise was to cost many a sailor's life for, even when people started to listen to Lind and accept what he said, citrus fruit was ignored in favour of useless doses of cheaper and more storable acids such as vinegar. Acid was acid, they reasoned.

THE LIMEYS

It was not until 1794 that the real hero stepped into the ring to conduct a twenty-three-week non-stop trial on board HMS *Suffolk*, bound for India. Gilbert Blane (1749–1834), later knighted for his reforms in navy hygiene, suspected that the now-dead Lind had in fact been on the right track with fresh citrus fruit but that the benefits lay in something other than citric acid; they had to be something else in the fruit that was responsible for averting the onset of scurvy. He was right, of course, but it would be another century before vitamins would be identified. Nevertheless, Blane made sure that every man got a daily dose of lemon juice to stay scurvy-free the whole trip. It was a few years before everyone accepted the prophylactic properties of citrus juice but, by 1800, every British ship carried barrels of lime juice, prompting American sailors to coin the still-popular nickname for anyone from these isles.

As more was learned about the malady and its prevention, more storable foodstuffs were favoured; some ships carried sauerkraut but most favoured was the growing of watercress from seed on wet blankets as required. Actually, citrus fruit is a remarkably poor source of vitamin C; limes, lemons and oranges yield 30, 40, 50mg per 100g respectively while watercress yields 662mg per 100g. Rosehips top the bill at 2,000mg per 100g.

AN UNBRIDLED SUCCESS

Scurvy is also responsible for the French predilection for horsemeat. During the Siege of Alexandria (1801), and again after the Battle of Eylau (1807), Napoleon's Surgeon-in-Chief, Baron Dominique-Jean Larrey (1766–1842), was confronted with serious outbreaks of scurvy and so, with no access at either time to any other source of vitamin C, he ordered the slaughter and cooking up of all spare horses to combat the pandemics. Meat does not have the same vitamin C content as some fruits and vegetables but the presence of amino acids facilitate a much higher absorption rate of the whole in the gut, which explains why the Inuit manages to stay healthy on a diet devoid of greens. Either way, this ration of horsemeat, seasoned with gunpowder and cooked over an open fire on the inside of a lancer's breastplate, saved the day and, when news of this reached home, the French thought it patriotic to follow suit.

Nitroglycerin

NITROGLYCERIN WAS FIRST SYNTHESIZED in 1847 by the Italian chemist Ascanio Sobrero (1812–88) while he was working under the aegis of the French chemist Theophile-Jules Pelouze (1807–67) at the University of Turin. In fact, Sobrero was so shocked by the power of his new explosive that he kept it to himself for a while and, when he did unveil it, he cautioned vigorously against its use in commercial blasting as it was just too unstable. But commerce ignored him; the new explosive made black-powder look like a child's toy and there were massive profits to be garnered from the contracts for roads, canals and railways to be blasted open. And, as Sobrero anticipated, the casualties in those blasting crews were so high that it started the race to find a safe form of nitroglycerin that would not capriciously explode at the slightest nudge. Enter the Nobels, father and son.

The Swedish engineer and entrepreneur, Immanuel Nobel (1801–72), was the man who invented plywood. He moved his family to Russia in 1838 where for the next twenty years they lived in St Petersburg under the patronage of Tsar Nicholas II, for whom they made weapons and, when Sobrero's cat was out of the bag, nitroglycerin. Unfortunately for the Nobel family fortunes, the Tsar died in 1855 and, with the Crimean War fizzling out the next year, they suddenly found their product-range redundant and the coffers empty.

By 1861 the Nobel family were back in Sweden with Immanuel and his third son Alfred (1833–96) still producing explosives but now from a small and very ramshackle factory on the outskirts of Stockholm. But this blew up, killing five workers, including Emil, Alfred's younger brother, and public pressure forced Immanuel to move his production to a barge on a lake outside the city.

Despite the time savings in large-scale engineering projects, the increasing death-toll in the labour forces caused a back lash of opinion against the use of nitroglycerin, which again threatened the Nobel family income.

AN EXPLOSIVE DISCOVERY

Alfred tried all sorts of adulterations, such as adding brick dust, to tame the beast but the only ones that worked seriously compromised the strength of the nitroglycerin to the point that one might as well be using the safer black-power it had replaced. Then it happened; the answer had been right under his nose all the time. An order of nitroglycerin was being prepared for shipment when Nobel noticed that one of the metal containers was leaking, with the nitroglycerin soaking into the *kieselguhr* – a cheap, porous silica-based mineral from northern Germany – that was packed between the containers to hold them still and protect them from buffeting. The resulting wet clay-like substance now resembled a kind of putty, which Nobel picked up and began to casually mould into shapes with his bare hands. When dried, the result was stable enough to be hit with a hammer – but don't try this at home – and, in small pieces, it could be set on fire without detonation. In fact, dynamite, as he would call it, was so stable that it needed a blasting cap to set it off.

DOUBLE TROUBLE

His next explosive discovery was born of the fortune that had previously smiled on John Wesley Hyatt in his search for synthetic billiard balls. Working in his lab one day in 1875,

Nobel, as had Hyatt, cut his hand on a broken lab-flask and, again as had Hyatt, he went to the cupboard to fetch down the bottle of collodion, this being a nitrocellulose-based liquid then used for dressings – the alcohol evaporates leaving a cellulose film over the wound, much as spray-on dressings do today. Unable to sleep for the pain in his hand, that night Nobel started to ponder, as one does, the possible results of combining the explosive nitrocellulose with nitroglycerin. He got up in the small hours and returned to the lab to try it out and, come sun-up, had produced a jelly-like substance that proved to be far more powerful than either of its two main ingredients, yet incredibly safe and stable without the 'sweating' problems of dynamite when it got a bit old. He called it gelignite.

A VIOLENT CURE

But there was more to nitroglycerin than death and destruction, with the first to actually try it out in the field of medicine being the American Constantine Hering (1800–80) who, in 1849, read Sobrero's papers that mentioned the terrible headaches afflicting all those who worked with the stuff. Being a homoeopathist and thus afflicted with the notion that like cures like, he started experimenting on those of his patients with persistent headaches and other chronic pain but all he managed to do was to give them more pain and tachycardia. Assorted homeopaths and a few mainstream medical figures played around with the substance throughout the 1850s–60s with all agreeing it had a pronounced effect on the body; it just needed someone to find the right use for it. This proved to be the English doctor William Murrell (1853–1912) who, in 1879, had absent-mindedly held the cork of his nitroglycerin bottle in his mouth while jotting down the need for more of the same on his laboratory 'shopping list'. He immediately experienced the kind of headache associated with tachycardia and a dramatic increase in the power of his heartbeat. He was, of course, aware that amyl-nitrate had, since 1844, been finding successes in the treatment of angina but he also knew that its use was off-set by some pretty unpleasant side effects, such as the involuntary relaxation of the body's sphincter muscles. That, along with its powerful vasodilatory action, explains why it is still sold in gay clubs and bars as the 'poppers' that heighten the enjoyment of certain activities, shall we say.

Murrell set up a small clinical trial to study the broader effects of nitroglycerin on angina patients with outstanding results and, more importantly, none of the aforementioned side effects; his only remaining problem being the explosive nature of the main ingredient. Working with his local pharmacist, Murrell eventually settled on the dissolving of the nitroglycerin in cocoa butter before pressing it out into pill form. This done, he spent an entire day stamping on them, hitting them with hammers on an anvil, throwing them out of the window, striking them like matches on a rough surface and finally tossing a few on the fire just for good measure.

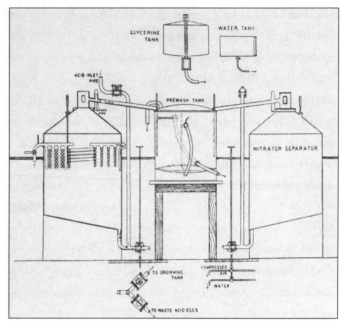

Diagram of a nitrating-house processing plant

BEHIND EVERY SUCCESSFUL MAN . . .

While all of this made Alfred Nobel one of the richest
men of his time, happiness always eluded him; he never
married and there were persistent but unproven rumours
of his having a gay romance with Sobrero. According to
Stockholm's Nobel Museum, the closest he came to a
relationship with a woman was when he ran the following
in newspaper personal columns at the age of forty-three:

'Wealthy highly-educated elderly gentleman seeks lady of mature age, versed in languages, as secretary and supervisor of household.' This brought forward the Countess Bertha von Suttner (1843–1914) who accepted the post of his secretary-housekeeper but, as a bit of a radical and a pacifist, she was so distressed at his business interests that she quit his Paris home two weeks later. Returning to her native Austria she rose to prominence in the pan-European pacifism movement, publishing *Lay Down Your Arms* in 1899, a book that was roundly praised for the depth of its perception and astute political pragmatism.

All that aside, she remained in friendly correspondence with Nobel up till his death and is generally credited with having put his mind on the track of the Nobel Prize scheme in the first place. Ironically, she would become the first female recipient when she was nominated for the 1905 Peace Prize.

In the end, Nobel became something of a recluse, deeply wounded by the general but unfair perception of him as some kind of mass-murderer. When his older brother Ludvig (1831–88) died, the press, asserting their usual masterly grasp on the wrong end of the stick, published notices of Alfred's passing, some of these sending him into a deeper depression. French papers were especially scathing in their trumpeting that 'The Merchant of Death is Dead' and castigating him for 'becoming rich by finding ways to kill more people faster than ever before'. Perhaps the final irony in the story is the fact that Nobel himself was prescribed oral nitroglycerin in the latter stages of his life, this being the only thing that kept him alive long enough for him to set up a new will, signed on 27 November 1895, and organize its implementation to lay the ground for the Nobel Prize: 'It sounds like an irony of fate that I should be ordered to take nitroglycerin internally. They call it Trinitrin so as not to scare the chemist and the public.'

A-ROUSING END

And it seems that the world is not yet done with Sobrero's brainchild. Condom manufacturers have launched a new line with the tip filled with a nitroglycerin-based gel called Zanifil. Despite this sounding more like a toilet-flush product, the manufacturers proudly announce this to be the way to go as the well-established vasodilatory properties of the nitroglycerin will not only have a 'viagra-like' effect but also better endow the wearer and prolong the intended activity. Perhaps even the morose Noble would have chuckled at that.

The Telephone

ALEXANDER BELL (1847–1922) – there was no 'Graham' in his name until the age of eleven – was first and foremost a teacher of sign language and other forms of communication to the deaf, and only got involved in the research that led to the telephone through his poor German translation of another speech therapist's work. Even when he did get involved in that research, his intention was not to produce an instrument of mass-communication but rather one that would allow a person to speak to one individual deaf person.

THE MOTHER OF INVENTION

Bell's father and paternal grandfather were both leading figures in the field of speech therapy and the teaching of systems enabling the communication to and from deaf and deaf-mute children. His own mother's deafness had a profound effect on the young Bell who was soon following in the family business. In 1863, his father took him to London to see a demonstration of automatons capable of rudimentary mechanical 'speech' with the star of the show an automaton built by Sir Charles Wheatstone (1802–75), who had in turn followed the work of the extraordinary Wolfgang von Kempelen (1734–1804), who had already stunned Europe with his other mechanical wonders, some of which were also capable of some sort of replicant speech. Bell was stunned; if mechanical speech was possible through mechanical resonance surely those same vibrations could somehow be transmitted to the inner ear of the deaf? He could actually talk to his mother and that idea held him in thrall from that night on.

LOST IN TRANSLATION

On returning home, Bell started to build his own artificial larynx and conduct experiments in the very nature of resonance and the transmission of sound in general.

Still only nineteen, he sent his findings in a paper to the English philologist Alexander John Ellis (1814–90), the man acknowledged by George Bernard Shaw as the model for Professor Henry Higgins in his *Pygmalion* (1912). It was Ellis's reply that inadvertently set Bell on the road to the invention of the telephone; his letter informed Bell that work in the field was already fairly advanced in Germany and enclosed a copy of *On the Sensations of Tone* by Hermann von Helmholtz (1821–94). This book would later be properly translated into English by Ellis but Bell, doing his best to cope with the original German, made a fortuitous slip up in his translation that led him to believe that the transmission of vowel and consonant through a mechanical medium was already possible. As Bell himself would later recount:

> Without knowing much about the subject, it seemed to me that if vowel sounds could be produced by electrical means, so could consonants, so could articulate speech . . . I thought that Helmholtz had done it . . . and that my failure was due only to my ignorance of electricity. It was a valuable blunder . . . If I had been able to read German in those days, I might never have commenced my experiments!

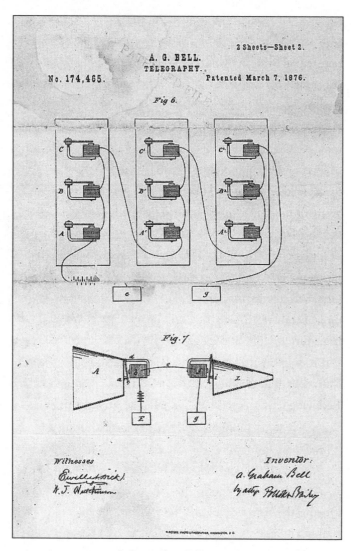

A diagram from Bell's patent

ADVANTAGE BELL

Spurred on by nothing more than his own bad translation, Bell continued his work and, by 1875, was in his own facility in Boston, Massachusetts, with the redoubtable Thomas Watson (1854–1934) at his side. They were experimenting with multi-tonal reeds connected to a wire and knew that the transmission of speech was within their grasp. But there was another player in the field who was even closer to that goal and Bell was well aware of the fact. Because of this, Bell had his lawyers staking out the patent office in Washington to tip him off in the event of Elisha Gray (1835–1901) turning up while he, Bell, was still trying to perfect his design. And that is exactly what happened; Gray walked into the United States Patent Office in Washington on 14 February 1876 to file patent on a telephone using a liquid transmitter while Bell is known to have still been in Boston. But it didn't matter that much as Bell and his patent-lawyer, Marcellus Bailey (1840–1921), had an ace in the hole: the Receiving Clerk, Zenas Fisk Wilber (dates unknown).

A PATENT PLOT

During the American Civil War, Wilber had served under Bailey and the two had enjoyed a shady relationship ever since Wilber had taken up his position in the Patent Office, perhaps helped into that very post by Bailey's machinations. By 1876, Wilber, by then a functioning alcoholic, was in hock up to his eyeballs to Bailey, who called in his marker by demanding that Wilber smuggle Elisha Gray's patent out of the office and meet up with him and Bell, who was already hot-foot for the city. Seeing his chance to wipe his slate clean, Wilber agreed and met the two conspirators and sat drinking as Bell modified his patent to incorporate the working bits of Gray's design that had eluded him. Bell gave Wilber $100 and Wilber made his way back to the office where he jiggled the paperwork to make it appear that Bell had filed his doctored patent before Gray, resulting in Bell being granted patent on 7 March 1876, the same day that Gray received notification that his application was denied.

Back in Boston, Bell spent days doctoring his notebooks and drawings to replicate Gray's design and, his equipment also so modified, he was ready for his famous demonstration of 'his' invention on 10 March, when he summoned his assistant from another part of the building, by phone, with the now-famous: 'Mr Watson, come here; I want you.'

Ten years later, on 6 April 1886, Wilber made a clean breast of the matter with a sworn affidavit confessing to his part in the theft and his payment by Bell and the cancellation of his debts by Bailey. Bell, of course, issued a vehement denial but by that time it was all but academic, with the Bell Telephone

Company a force to be reckoned with. It would be 1990 before Bell's doctored notes and drawings were made public and the scandal brought to wider audience.

Lobotomies

ON THE MORNING of 13 September 1848, the 25-year-old Phineas Gage (1823–60) set out to work on the Rutland and Burlington Railroad, blissfully unaware that he was about to make significant advances in the field of neurosurgery – albeit by painful accident – and pave the way for one of the most despised medical procedures of all time: the lobotomy.

LIKE A HOLE IN THE HEAD

Gage was the foreman of the blasting team that was working to the south of Cavendish, Vermont, where they were clearing an inconvenient outcrop of rock that stood in the way of the advancing track layers. With the time approaching 4.30 p.m., Gage himself was preparing the last batch of charges of the day and had begun packing the bored holes with the priming charges when it happened. There must have been a spark from his tamping iron striking perhaps flint in the borehole, because the primer-charge detonated and fired the iron out of the hole like a bullet. It went straight through Gage's head, to land some eighty feet from the scene. The forty-three-inch-long iron, blunt on the tamping end and tapered at the other, had entered the left side of Gage's face and exited to the right-hand side of the crown of his head. More extraordinary than the fact that Gage was not stone dead was the fact that he did not even lose consciousness. After shaking what was left of his head, he walked unaided to a wagon and sat up chatting all the way to the nearest doctor in Cavendish, which was a very bumpy mile away.

The first physician to see Gage, Dr Edward H. Williams, clearly did not believe his patient's tale:

I first noticed the wound upon the head before I alighted from my carriage, the pulsations of the brain being very distinct. Mr. Gage, during the time I was examining this wound, was relating the manner in which he was injured to the bystanders. I did not believe Mr. Gage's statement at that time, but thought he was deceived. Mr. Gage persisted in saying that the bar went through his head. Mr. G. got up and vomited; the effort of vomiting pressed out about half a teacupful of the brain, which fell upon the floor.

KEEPING HIS HEAD

News of Gage and his remarkable survival spread through the town like wildfire, attracting the attention of another physician, Dr John Martyn Harlow (1819–1907), who would sustain contact with Gage until his death from seizures in 1860; he would also document the rest of Gage's life in what is recognized as one of the longest patient follow-up programmes in medical history. Writing of his first contact with the seemingly untraumatized Gage in the *Boston Medical and Surgical Journal* later that same year, Harlow observed:

> You will excuse me for remarking here, that the picture presented was, to one unaccustomed to military surgery, truly terrific [here the doctor is using the term in its original and proper sense of terrifying, not enjoyable]; but the patient bore his sufferings with the most heroic firmness. He recognized me at once, and said he hoped he was not much hurt. He seemed to be perfectly conscious, but was getting exhausted from the haemorrhage. Pulse 60 and regular. His person, and the bed on which he was laid, were literally one gore of blood.

Harlow also recorded that the tamping iron had:

Entered the cranium, passing through the anterior left lobe of the cerebrum, and made its exit in the medial line, at the junction of the coronal and sagittal sutures, lacerating the longitudinal sinus, fracturing the parietal and frontal bones extensively, breaking up considerable portions of the brain, and protruding the globe of the left eye from its socket, by nearly half its diameter.

Gage remained under Harlow's care for some weeks, with the doctor seeing him through a series of infections and setbacks but, in the November of that year, Gage was well enough to return to his family in New Hampshire where he achieved a general improvement and final recovery. The next April he returned to Cavendish to see Dr Harlow who observed on the top of Gage's head:

. . . a deep depression, two inches by one and one-half inches wide, beneath which the pulsations of the brain can be perceived. Partial paralysis of the left side of the face . . . his physical health is good, and I am inclined to say he has recovered. Has no pain in head, but says it has a queer feeling which he is not able to describe.

Basically, the poor chap had given himself a violent pre-frontal lobotomy so he was entitled to feel a bit queer from time to time. But the important thing as far as we are concerned here is the fact that to one degree or another Gage had indeed changed; he was no longer of the same character.

FALSE ACCUSATIONS

Gage's character certainly changed, with him becoming more placid and introverted than his work colleagues remembered him. There are lurid tales of the incident having turned him into a drunken and abusive blasphemer who was much given to sexually inappropriate behaviour in public, but all these come from the pens of people who had never met him. Many of these accounts accuse him, for example, of being violent and abusive towards his cowering wife and children, whereas in fact he had neither. True, the railway did not hire him again but, by the time Gage was again fit for work, all the blasting in that section had finished. Far from being unable to find work, Gage made a bit of money by meeting and greeting visitors to P.T. Barnam's American Museum in New York, before travelling to Chile to work as a stagecoach driver on the Valparaiso to Santiago run. This he did for several years and no such job would be long held by an abusive drunk given to self-exposure.

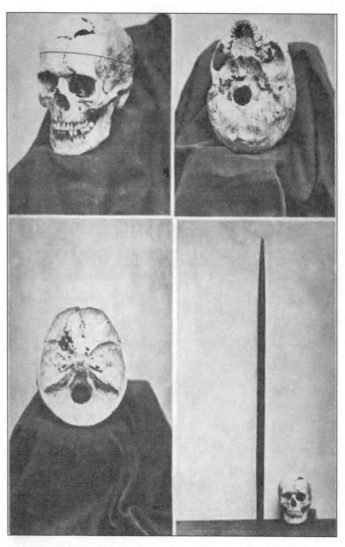

Multiple views of Phineas Gage's exhumed skull, with tamping iron

Eventually, in 1859, Gage took a turn for the worse and returned to the USA to stay with his mother, then living in San Francisco. After a brief but rapid decline, involving seizures, Gage died – but with his real legacy yet to unfold.

ON THE BRAIN

Gage became a benchmark case in neurology and a standard reference for anyone specializing in anything to do with the brain, as his case proved, to one degree or another, the personality resided in the prefrontal lobes of the brain. The year after Gage's death, French surgeon–anthropologist Paul Broca (1824–80), announced that the prefrontal lobes must be the seat of whatever it is that marks us out from the animal kingdom, this prompting hoards of dilettante brain surgeons to start capriciously hacking away at the brains of assorted animals. But the first to try and replicate Gage's injuries in humans was Gottlieb Burckhardt (1836–1907) who got stuck into the heads of six inmates of the small asylum he ran in Neuchâtel in Switzerland. Undeterred by the fact that two of these unfortunate guinea pigs died a couple of days later, Burckhardt presented the results, claiming the surviving four to be either cured or greatly improved. Fortunately for his other patients, the reception he received ranged from the sceptical to the openly hostile and, roundly castigated and marginalized, Burckhardt mooched off into the obscurity he so richly deserved. But others were already reaching for their hammers and chisels as the bandwagon that Gage had started was slowly picking up speed.

MONKEY BUSINESS

In 1935, a sixty-one-year-old brain surgeon and ex-politician called Egas Moniz (1874–1955) had travelled from his native Portugal to attend the Second International Neurological Congress in London, where he listened intently to a couple of Americans from Yale University. Physiologist John Farquhar Fulton (1899–1960) and animal physiologist Carlyle Jacobsen (dates unknown) were both highly familiar with the Gage case, and had conducted frontal lobotomies on two of their experimental chimpanzees, in which they had surgically isolated all the connections between the prefrontal lobes and the rest of the brain. Although the two animals, Becky and Lucy, remained friendly and alert, they were no longer given to experimental neurosis or frustration at any failure in the tasks set them. In fact, both animals had lost their ambition, their drive to succeed and much of their puzzle-solving ability – but that didn't frustrate them as they simply couldn't give a monkey's about anything by then. However, Fulton and Jacobsen rightly thought it unwise to dwell on this aspect, focusing their talk instead on the fact that both subjects were now placid and resigned to their captive environment.

THE GOOD DOCTOR

Moniz scuttled back to Lisbon where, assisted by Pedro Almeida Lima (1903–85), he tried out the Americans' technique on a female patient on 12 November that same year. After drilling holes in her head, Moniz injected areas of the prefrontal lobes with novocain and then alcohol to destroy the white fibres that connect the lobes to the rest of the brain. Proclaiming this first operation a success, Moniz decided to abandon the injection method and go for the knife in an operation he termed the leucotomy, or 'cutting of the white matter'. Moniz and Lima preyed on agitated depressives, as these *appeared* to 'benefit' most from the procedure. Also, having once been a political animal, Moniz was skilled in the art of presenting the unpalatable in such a way as to make it attractive, so he was always sure to refer to his butchery as psychosurgery.

Next, he hacked away at the brains of assorted schizophrenics, all of whom he pronounced to be greatly improved. In fact, they were all sluggish, moribund and disorientated. Moniz was a liar; he either skewed the figures or lied about the true results and hid the deaths. 'Prefrontal leukotomy is a simple operation, always safe, which may prove to be an effective surgical treatment in certain cases of mental disorder'. How did he know? He never followed up any of his victims to see how they were coping; many had died of cerebral haemorrhage while others were reduced to vegetative states. The most rational criticism of his 'work' came from one of his own patients who walked up and shot him in 1939; Moniz was hospitalized and confined to a wheelchair for a while but was soon well enough to continue the 'good work'.

ALICE IN LA-LA LAND

Unfortunately for several thousand Americans, Dr Walter Jackson Freeman II (1895–1972) of St Elizabeth's Hospital in Washington DC became apprised of the alleged benefits of Moniz's procedures and decided to get in on the act. Trying his hand on a few of the residents of the basement morgue who were beyond complaining about the fact that he was wholly untrained as a surgeon of any kind, Freeman decided the technique was a piece of cake and opened for business. His first victim, Alice Hammatt of Topeka, Kansas, went under his untrained knife on 14 September 1936, Freeman having bullied her and her husband into signing the consent forms by telling them it was a straight choice between the operation and her being locked up in an asylum for the rest of her life. Alice was, it seems, most worried about losing her luxuriant curly hair in the operation, which Freeman assured her would not be the case. As things turned out he was lying but, after she came round with a shaved head, Freeman was pleased to note that Alice didn't seem to care one way or the other; she just shuffled off home.

PICKING PEOPLE'S BRAINS

In 1945, Freeman had heard of an exciting new technique developed in Italy where, bored with drilling access holes in his patients, Amarro Fiamberti (dates unknown) was going in through the top of the eye socket with an ice pick. Having stabbed a few grapefruits in his kitchen at home before returning to his most uncomplaining of patients – the corpses in the morgue – Freeman felt ready to give it a go. It was swift; it could be done in the office without resort to general anaesthetic, the patients could just go home in a taxi and, best of all, it was flamboyant, shocking and dramatic – and Freeman loved to put on a show. He frequently invited the press to demonstrations and 'lobothons' during which he romped through as many as twenty or so patients, spending as little as ten minutes on each, from the rendering of the patient unconscious with electroconvulsive shock to the finish. His favourite 'party-trick' on such occasions was the double-handed gambit, which had him wielding two ice picks and going in through both eye-sockets simultaneously while joking about tossing salad or spaghetti in a bowl as he stirred about. As it happens, Freeman's first two ice pick patients of 1946 were Sallie Ellen Ionesco (1917–2007) and Helen Mortensen (1915–67), this latter turning out to be his nemesis.

STEALING HER YOUTH

But the really dark side of the increasingly popular operation was the fact that it was not just the mentally ill who went under the knife or ice pick. In America, homosexuality was widely believed a mental illness at that time, so countless men and women had their 'abnormal urges' curbed by the lobotomy, as indeed did others who were simply deemed 'troublesome' or wilful by their parents, the most famous such case being that of Rose Marie 'Rosemary' Kennedy (1918–2005), the eldest daughter of Joe Kennedy (1888–1969). To be fair, the girl could be a bit of a handful and was showing signs of wilful promiscuity, leaving Joe fearful of some social or sexual scandal, and this from the man who bootlegged booze during the Prohibition, conducted an open affair with actress Gloria Swanson and raised more than a few eyebrows with his support for Hitler.

Without telling his wife or any other members of the family, Joe Kennedy quietly had Rosemary declared mentally ill and in need of 'calming down'.

Well, Rosemary was certainly more manageable after Freeman had finished with her; so manageable in fact that she just sat rumpled in a wheelchair, institutionalized for the rest of her days. She was twenty-three at the time but that didn't bother the man who had operated on over twenty pre-teen children. In Freeman's own words:

We went through the top of the head, I think she was awake. She had a mild tranquilizer. I made a surgical incision in the brain through the skull. It was near the front. It was on both sides. We just made a small incision, no more than an inch. The instrument Dr Watts used looked like a butter knife. He swung it up and down to cut brain tissue.

Every now and then the cutting stopped while Freeman asked Rosemary questions or asked her to sing a particular well-known song: 'We made an estimate on how far to cut based on how she responded'. In a nutshell, they kept cutting until she became incoherent.

Hidden from public view, the shell of Rosemary had few visitors but her sister Eunice was a regular and, perhaps inspired by her sister's fate, she was fundamental to the founding in 1968 of the Special Olympics.

BACK FOR MORTENSEN

Eventually, however, time was called on Freeman's lobotomies by the death of the aforementioned Helen Mortensen. Obviously not satisfied with the results of her first encounter in 1946, Helen was back for more in 1956 and again in 1967, after which her brain must have been in a sorry state to say the least. This time the ice pick severed an artery in her brain and she died, with the American Medical Association finally waking up to the horrors of the Freeman circus, banning him from performing any surgical procedures of any kind.

By that time Freeman had butchered the brains of about 3,500 of his compatriots with one of the youngest survivors being twelve-year-old Howard Dully (b. 1948), who Freeman diagnosed as 'wilfully disobedient'. It took the poor chap decades to recover from that 'surgery' but he is at last on an even keel and happily settled in San Jose, California, where he works as a tour bus operator. But he is one of the lucky ones; in all, over 40,000 Americans were lobotomized, as compared with only 10,000 across Western Europe. Although a popular fallacy has long held that Swedes are more given to depression and suicide than any other race, it must be said that between 1944 and 1966 they lobotomized over 4,500, which is nearly three time the per capita score of any other nation held in the grip of this medical nightmare – who knows, perhaps that very statistic is the origin of the aforementioned myth?

OTHER NOTABLE CASES

No one was safe. Actor Warner Baxter (1889–1951), better known to older readers as *The Cisco Kid*, was advised in 1951 to have a lobotomy to help manage the pain of his arthritis –and he never felt anything at all after that. Rose Isabel Williams (1909–96), who was admittedly given to bouts of violent and promiscuous behaviour, was operated on in 1943 and settled down to quietly and obsessively collect little glass animals, thus prompting her brother, Tennessee, to write *The Glass Menagerie* (1944).

Thalidomide

OTHER CHANCE DISCOVERIES hide a similar dark story, such as that of thalidomide. The usual legend tells of the drug first being isolated in the laboratories of the German pharmaceutical company Grünenthal, which, ironically, translates as '(pleasant) green valley'. The official company line has it that, in 1953, while looking for a way to synthesize antibiotics from peptides, their Head of Research, Herr Doktor Heinrich Muückter (1914–87), stumbled across the drug, which seemed to have an unusually calming effect on his test animals.

SLEEPING KILLER

With no idea of the horrors they were about to unleash, the company marketed it as Contergan in 1954, an over-the-counter sedative. Desperate for a bestseller, Grünenthal flooded Germany with free samples and, by 1961, it was a leading drug in over forty-six other countries. How many of those early users were, by coincidence, pregnant woman is not known but the real damage occurred between 1957 and 1962, when it was specifically targeted at such patients as a 'completely safe' answer to morning sickness.

Before that targeting, Grünenthal employees were themselves using the drug, with the first evidence emerging on Christmas Day 1956, when a lab assistant's wife gave birth to a baby with no ears, either external or internal. This was quickly hushed up but between 1959 and 1961 a further eight employee families of Grünenthal endured births of terrible deformity. By this time, everyone within the company knew full well the dangers and was also aware of the rising tide of evidence reported from around the world. Not one single Grünenthal employee would either take the drug themselves or allow their families to touch it. But the people at the helm of Grünenthal kept their shoulders to the floodgate of evidence and warnings while harvesting as much profit from its horror as possible before it was finally forced off the market in November 1961.

In 1961, for example, Herr Doktor Mückter, whose base-line salary was 14,400 DM, was paid a total of 320,000 DM in bonuses that year alone. At today's value that is over £1 million, so the profits reaped from such widespread deformities and the attendant misery would have been huge. Not until September 2012 did the company issue any kind of apology and acknowledgement of responsibility, which most of the victims, as one might expect, dismissed as too little, too late.

SARIN AND THE RING OF POWER

But it now seems that the drug predates even the setting up of Grünenthal in 1946 and that its accidental discovery occurred in nerve-gas testing in a different environment altogether, and one that was as far away from any Peaceful Green Valley as it was possible to get.

The nerve-agent Sarin had been developed at IG Farben in 1938 by Gerhard Schrader (1903–90) while he was trying to find an answer to Third World famine. His co-researcher on that project was Otto Ambros (1901–90), who suggested naming their discovery from the surnames of the prominent team-members Schrader, Ambros, Rudiger and LINde. Sarin was seen by Schrader, at least, as an organophosphate insecticide that would wipe out locusts and other parasites that plagued the Third World. But Ambros, already a closet member of the Nazi

Party, had other ideas, and reported his thoughts for 'other uses' to his and IG Farben's 'puppet-masters'.

HORRORS OF WAR

Throughout the war, Ambros was ostensibly attached to the IG Farben slave-labour factory at Monowitz Concentration Camp, which produced synthetic oils and rubber with the help of over 83,000 slaves drafted in from neighbouring Auschwitz. The labour force was inspected daily at the gate, with those too weak to be useful returned to their fate (but, as IG Farben also held the patent for the Zyklon B that would be used in such extermination, the company saw a profit whichever way it went).

Ambros spent much time in Auschwitz where he selected the fittest of the inmates to be exposed to Sarin in a variety of ways so he could gauge the effectiveness of his experimental antidotes, one of which he would be peddling as the sedative Contergan in less than fifteen years. This, irrespective of the fact that he had seen the side effects on his camp guinea pigs, some of whom had been pregnant at the time of these tests. Even male recipients developed peripheral neuropathy, severe damage to the nerve pathways at the extremities, as indeed was the case with many non-pregnant recipients of Contergan. After the war, along with several other senior IG Farben figures, Ambros was prosecuted for war crimes at Nuremberg;

Otto Ambros on trial in Nuremberg over his work on chemical warfare agents

Ambros was sentenced to eight years and the industrial giant was dismantled back to the individual companies that had clustered to form the monster in the December of 1925 – Agfa, Bayer, Hoechst and Cassella.

SINISTER SCIENCE

As Ambros began his sentence, Hermann Wirtz (1896–1973), an enthusiastic Nazi and devoted party member, was busy setting up Grünenthal in Aachen, already a favourite gathering place for ex-Nazis and a staging post on one of the major rat-lines helping the worst of their kind escape to South America.

During the war, Wirtz's firm had concentrated in the main on the production of soaps, household products and fragrances and had benefited greatly from the Ayranization Programme, a euphemism for the process of stealing firms from Jewish owners and giving them over to good party members.

It is probably fair to say that all sorts of companies hired ex-Nazis, but Grünenthal seems to have attracted far more than its fair share. Among others, there was Doktor Martin Staemmler (1890–1974) who, during the war, had been Secretary of the Race Policy Office and had written much on the subject of racial hygiene and how to best achieve it. There was also SS-Doktor-Colonel Ernst-Günter Schenck (1904–98) who spent the war conducting some very strange dietary experiments in Dachau and Mauthausen camps, where he was trying to invent some new super-ration for the SS troops; 370 inmates died in one of his trials alone. He was also one of the few attendees at the reception after the mercifully short-lived marriage of Hitler and Eva Braun, so hardly a peripheral player in the hideous circus. Another leading light of Grünenthal was the charming Heinz Baumkötter (1912–2001), Doktor-Hauptsturmführer in charge of execution selection at Mauthausen and Sachsenhausen Concentration Camps and, finally, Doktor Heinrich Mückter who had spend much of the war conducting nightmare experiments on the inmates of Auschwitz and Buchenwald camps in an effort to find a new cure for typhus. And, as soon as Ambros was released in late

1952, he was hot-foot down to Aachen to join this coterie of medical philanthropists, just in time for the announcement of the 'new' wonder-drug.

The main point here is that we are not looking at a group of men who would be in the least bothered about causing the 90,000 known miscarriages attributed to thalidomide/Contergan or the deformed births of the unlucky ones if there was a profit to be had.

TOO JUNG TO KNOW

When Grünenthal first registered its patent in 1954 there were clear indications that tests on humans had already been conducted but even the monsters within the company knew that these were hardly the sorts of trials they could boast of or dare make public. So, they had to hurriedly organize some favourable reports. First to be drawn into this scheme was a tame doctor they already had on retainer, a Dr Jung of Cologne, who ran some haphazard tests on about twenty patients for less than four weeks before waxing enthusiastic about the drug's myriad benefits. He had administered it to four youths who were suffering moral anxiety over their excessive masturbation; they persisted with their excessive and solitary pleasures but no longer felt any pangs or remorse for so doing. Jung also claimed it cured premature ejaculation in a sample of his married patients whose marriages had been saved as a result; on the basis of these shaky and unsubstantiated reports, Jung stated in June 1955 that thalidomide was indeed a wonder-drug that was ready for market.

FUTURE ENDORSEMENTS

Various drugs drift in and out of fashion or favour, frequently being shelved, only for someone to find a new use for them in an ever-changing market. And such is the case with thalidomide, for which some see a brave new future in the treatment of leprosy. In 1964, Jacob Sheskin (1914–99) of Jerusalem's Hansen Leper Hospital had given a male patient a hefty injection of thalidomide to fight the infection, leaving the man unable to sleep due to certain adverse effects. In the morning, Sheskin was staggered to see his patient up and about and dramatically improved. As it turned out, it was the drug's ability to restrict the development of blood vessels, this being responsible for the deformities and withered or non-existent limbs, which works in others' favour by restricting the abnormal growths and contusions associated with leprosy. Concentrating on this action of the drug, scientists are conducting thalidomide trials against arthritis, lupus, HIV and cancer, but only with male patients or such female patients who have been very carefully selected and screened.

STRINGENT TRIALS

Another tame doctor, and another who had never before conducted pre-launch trials on any drug of any description, Dr Konrad Lang, was drafted in to the fiasco and paid handsomely for his biased opinions. Ignoring all the accepted protocols and resorting to tactics worthy of any of the ex-Camp doctors paying his fee, Lang selected a batch of forty mentally impaired children in the University Clinic at Bonn where he administered twenty times the adult dose over a nine-week period without telling anyone what he was doing, especially not the parents. His guinea pigs ranged in age from a few months to five and six, and despite one of the group dying of circulatory collapse, two others of heart failure, a one-month-old baby going into convulsions and losing her sight, and another three-month-old baby dying of heart failure, Lang reported back to his Grünenthal paymasters that: 'In general terms, Contergan could be described as a rapid-acting sedative particularly suited for use with children.'

Come the day of reckoning, Grünenthal claimed to have lost all the paperwork relating to these and other 'stringent trials'.

Radiation Exposure

THROUGHOUT THE 1950s, the Nevada Desert played reluctant host to Operation Plumbbob, which was the umbrella name for countless nuclear tests, one of which, codenamed Harry and detonated on 19 May 1953, accidentally killed off actor John Wayne (1907–79) and half of the cast and crew of his most disastrous movie, *The Conqueror* (1956). Most nominated film for the Golden Turkey Awards, John Wayne foolishly agreed to play Genghis Khan and, to the amusement of all, did so in cowboy-mode while strutting around in box-pleat trousers and a fur hat. But, behind the laughter, he was about to provide America with its own study group to establish once and for all the link between exposure to radiation and cancer.

IN THE DARK

It may seem hard to countenance today, but in the 1950s no one thought the fallout risks from nuclear tests to be that significant; it was just a big bang, wasn't it? It must be remembered that it was still only eight years on from Hiroshima and Nagasaki, and cancers need time to develop. The first murmurings of a link were not even raised in Japan until the mid-late 1950s and even then few accepted the findings, insisting instead on more extended research. Many of those exposed had moved to other countries, died of other causes or otherwise dropped off the radar. Others, just to cloud the issue further, recovered from their external burns and injuries to live into their eighties and nineties. Factor in that a given percentage of any population, unexposed to high levels of radiation, is going to develop a range of cancers and the problem of establishing a cause-and-effect chain of evidence becomes clear. It is a dramatically different proposition to establishing a causal link to a short-termed and contained effect in a static population.

RADIATION FOR SALE

Even Marie Curie (1867–1934), who died not of cancer, as many believe, but from prolonged radiation exposure, did not understand the dangers; she carried samples around in her pockets and noted with some amusement how they glowed in the dark. All of her papers and personal possessions are currently stored in lead-lined boxes and those wishing to see them have to first suit-up, an indication of how far the understanding has shifted.

Even into the 1960s, radiation was thought to be beneficial in everyday life – even fun! The big health-craze of 1930s America was radium waters; the market-leader, Radithor, containing a staggering 2 microcuries, one of radium 226 and another of radium 228. Until the late 1950s, this bizarre market brimmed over with such delights as radioactive cosmetics, toothpaste, suppositories, condoms – bet those worked a treat – tea, chocolate and even the Atomic Energy Lab Kit, as marketed in 1950 by A.C. Gilbert & Company, this being the toy for the boy with everything. Costing about $500 at today's values, the kit contained numerous radioactive samples and instruction on how to build your own atomic cloud chamber. And what of those fashionable wristwatches, popular until the late 1960s, with the hands and numeration that glowed in the dark so you could tell the time at night? And how many readers remember the fluoroscopic X-ray machines that were, until the early 1960s, standard equipment in all high-street shoe shops? Intended to show the wearer their feet inside the shoe to ensure a good fit, not only were customers happy to make frequent use of such devices that would now be considered bordering on the criminal, but mothers were happy to let their children play with them, X-raying assorted parts of their bodies for almost suicidal lengths of time. How would one find out how many of them later developed cancers?

ATOMIC TOURISM

At the time of those Nevada tests there was no cover-up (that would come later) or dark conspiracy theory; had it been known of the link to cancer then someone in the atomic fraternity would have called out – but even the people involved in the tests themselves did not take any of the precautions that would be standard today. Military personnel took 'shelter' in trenches and put their hands over their eyes if facing the flash, while the technical staff wandered around in shorts and white coats.

Funded by Howard Hughes (1905–76) – who made more turkeys than most – the cast and production crew of *The Conqueror* descended on the town of St George, Utah, in 1954, bringing a touch of glamour and some very welcome dollars to the local economy.

Meanwhile, about 100 miles away at Yucca Flats in neighbouring Nevada, the United States Atomic Energy Commission (AEC) was setting off one device after another – a staggering 827 in all – telling the people who would later become collectively known as the Downwinders, that all was perfectly safe and that, no, the tests had nothing to do with all the dead sheep and cattle in the area. In fact, the AEC leafleted the town telling the residents that, among other things, they should be proud of the fact that: 'You are in a very real sense active participants in the nation's atomic test programme.'

Thus unconcerned, many of the locals often went out into the desert in mass picnics to watch the fun. Also, in the May of 1953, the senior year of Middle Park High from Granby, Colorado were visiting the area and were invited by AEC officials to get up early so they too could come out and watch the test and tell all their friends back home what they had seen. There was no secret and, unfortunately, no fear.

WAKE-UP CALL

But then the patterns began to emerge, not just in the locals but also in that cluster of 220 famous visitors, of whom ninety-one developed assorted and obscure cancers. (Later tracking of known visitors to the film set, including the stars' sons, daughters and partners, produced equally alarming results.) Although there had been no active tests while the film crew was in Utah, there had been eleven in the preceding twelve months, with 'Dirty' Harry producing over twice the fallout of Hiroshima and Nagasaki together, which ensured that the ground they filmed on, the dust they inhaled, the water they drank and the very air they breathed was all as contaminated as the locally slaughtered beef they consumed with relish.

Just to make matters worse, about sixty tons of that contaminated desert sand was brought back to the Hollywood studios to ensure colour-continuity in any retakes or additional scenes. Dr Robert Pendleton, Professor of Biology at the University of Utah, stated in 1980:

> With these numbers, this case could qualify as an epidemic. The connection between fallout radiation and cancer in individual cases has been practically impossible to prove conclusively. But in a group this size you'd expect only thirty-some cancers to develop. With ninety-one, I think the tie-in to their exposure on the set of *The Conqueror* would hold up in a court of law.

But, with the realization of the by now undeniable cause-and-effect, the cover-up and weasel tactics began.

TOO LITTLE, TOO LATE

With increasingly stark evidential studies coming out of Japan to mirror the domestic findings, successive American administrations found themselves besieged by representations for apologies and compensation from the Downwinders

– 15,000 of whom had developed cancers, including Scott M. Matheson (1929–90), Governor of Utah, who lost ten family members to cancer before dying himself of multiple myeloma, a very rare cancer, at the age of sixty-one. Delaying as long as was humanly possible to give as many of the group the chance of quietly dying to reduce the inconvenience, the Bush Administration of 1990 finally passed the half-hearted Radiation Exposure Compensation Act under which people who could prove a link between their exposure and certain conditions might receive $50,000, with uranium miners of the area being entitled to $75,000. The trouble was, most of those miners were Navajo who had only gone through tribal nuptials with no paperwork.

SPACED OUT

And so to Pascal B, the Plumbbob test which accidentally put the first man-made object into space – maybe. On the 27 August 1957, Test Director Dr Robert R. Brownlee was overseeing an underground test that should have been of an extremely limited yield.

Pascal B was to be conducted in a 500-foot deep shaft, so the first milliseconds of the blast could be filmed. The shaft was about four-feet wide, lined with concrete and with a five-feet thick collimator, or plug, with a tiny hole at the centre for the camera to do its very short-lived work. And, finally, there was a four-inch thick manhole cover welded to the top of the chamber. That cover would have weighed in the region of about 900kg and, according to the calculations of some, would have sailed serenely past Pluto a few years ago.

A balloon shot test firing during Operation Plumbbob

Something went wrong with Pascal B and it blew at several thousand times the anticipated yield, perhaps 50 kilotons, turning the capped shaft into a sort of atomic gun with the manhole cover as its only bullet. Escape-velocity calls for an object to travel at a minimum of 11km/sec and the external high-speed cameras over the shaft only capture the cover in flight on a single frame, which, giving some indication of its speed, left Brownlee to reckon that it was travelling at about 67 km/sec. To put that in terrestrial terms, it would have gone from one side of Australia to the other in less than a minute.

There are those who say that just because the cover was never found does not mean it escaped the earth and went on its merry way through space, besides, surely it would have vaporized? Possibly; but others point out that it would, under those circumstances, have been protected in a bubble of super-heated air until it escaped the earth's gravity.

The Cellphone

THE BIRTH of the technology that made possible the mobile phone was the product of a chance meeting and conversations between a 1930s screen siren, an Italian submarine commander of the First World War and a decidedly camp and avant-garde American musician. Friedrich Mandl (1900–77) was one of the biggest arms dealers of the interwar years and, as such, he managed to secure a patent-franchise to manufacture the British Whitehead torpedo. An Austrian Jew of considerable wealth, he had no qualms over dealing munitions to both Hitler and Mussolini – nor in later life supporting the Peron regime – and it seems that his stunning and voraciously bisexual wife was frequently slipped into a prospect's bed to seal a deal. Born Hedwig Eve Maria Kiesler (1914–2000), this was the lady who would later crop up in Hollywood as

Hedy Lamarr, after growing weary of being pimped out by her controlling husband who made her the only woman in history to have slept with both Hitler and Mussolini.

But there was a lot more to Lamarr than her stunning looks; she had brains too. Despite being a mere nineteen at the time of her marriage to Mandl, she always paid avid attention during the talks she was obliged to adorn, and thus acquired considerable knowledge of the operating systems of the Whitehead torpedo. These were still sharp in her mind when she bumped into the second of our three serendipitous players at another Hollywood party – the Italian submarine commander, retired.

TORPEDO-RE-MI

Our Italian submarine commander had married into the Whitehead torpedo dynasty, marrying Robert Whitehead's granddaughter, Agathe (1891–1922). After the war, Agathe and her husband enjoyed a life of quiet domestic bliss in Austria until scarlet fever carried her off, leaving her husband to cope with such a brood of children that he was obliged to hire in someone called Maria from the local nunnery. This man was, of course, Ritter (Sir) Georg von Trapp (the film promoted him to Baron) who, with those same kids, would later warble his way to fame in America.

The von Trapps quit Europe for America – leaving openly by train and boat, not over the mountains pursued by the Nazis – where Georg ended up in conversation with Hedy Lamarr. Having little else in common, that conversation soon settled on their swapping notes and anecdotes about the Whitehead torpedo, in particular the vulnerability of the radio-guidance system to jamming by the target vessel which could quickly lock on to whatever frequency was guiding the in-bound torpedo and send it awry with a jammer.

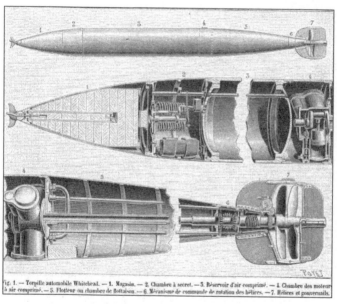

Fig. 1. — Torpille automobile Whitehead. — 1. Magasin. — 2. Chambre à secret. — 3. Réservoir d'air comprimé. — 4. Chambre des moteurs à air comprimé. — 5. Flotteur ou chambre de flottaison. — 6. Mécanisme de commande de rotation des hélices. — 7. Hélices et gouvernails.

A 1891 diagram of a Whitehead torpedo mechanism

TARIFFS AND CONTRACTS

In *Brando Unzipped* (2005) by Darwin Porter, when the subject of the biography recalled asking Hedy Lamarr at a party if this was true. She responded:

'I was married to Fritz Mandl . . . if he ordered me to sleep with one of those dictators I did the bidding of my husband. That way, he could get fatter contracts. Hitler was all posturing and was not a man; Mussolini was the most pompous ass I've ever known. Imagine stopping every minute to ask how he was doing?'

BOOBS AND THE BAD BOY

And that might have been that, had Lamarr not fancied having bigger breasts. Within a couple of days of this conversation, Lamarr was pounding on the door of one of America's self-proclaimed 'bad-boys of American music', the extremely 'flamboyant' and avant-garde composer-director, George Antheil (1900–59). For reasons best known to himself – he had no training in the field – Antheil fancied himself an expert in female endocrinology and had written articles on the subject, these appearing in such cutting-edge medical publications as *Esquire*. He also wrote *The Glandbook for the Questioning Male*, a sexual predator's guide to reading the supposed signs of 'sexual availability and susceptibility' of women which, he claimed, were betrayed by their complexion as generated by their hormonal balance. Antheil also promoted himself as a woman's hormonal guru and Lamarr went to see him to get his advice on how she could become bigger up top, these being the days before cosmetic breast surgery.

THE SOUND OF MUSIC

Soon realizing that Antheil didn't know one end of a woman's anatomy from the other, Lamarr shifted the conversation to the next work of art he was about to inflict on the ears of his fellow countrymen – *The Ballet Mechanique*. A discordant cacophony of pretentious tripe, this called for a number of pianos and other instruments to play in synchronized fashion and the pair began discussing how this would be best achieved. Suddenly, it all fell into place; the previous conversation with von Trapp and Antheil's problem meshed into an idea: why not have the torpedo guidance system running on a frequency-hopping device with, say, eighty-eight different frequencies – the number of keys on each of the pianos they were trying to synchronize? No enemy ship would stand a chance of fixing on a frequency that would itself change a second later, and keep doing so until the torpedo slammed home. On the 11 August 1942 their perfected system was granted U. Patent 2,292,387 and Lamarr approached the US Navy to explain how even a plane could drop such a torpedo in the vicinity of the target and then sit back and steer it home.

THE TOO-FEMALE BRAIN

For the first time in Lamarr's life her beauty worked against her. Basically the navy thought she was far too pretty to be taken seriously and, after making facetious comments about piano-playing torpedoes, suggested she should stick to selling war-bonds and kisses and leave the big boys to sort their toys. Despite her explaining how the device could be made small enough to fit inside a wristwatch and how torpedoes could be fired from a surface vessel that could then speed away to safety and leave the guidance to a high-altitude plane, she was patted on the head and sent packing.

They were the fools and she was years ahead of her time. After the war, someone in the American War Department remembered her idea and, swapping ticker-tape style roll for electronic circuitry, put her ideas to use in the 1962 blockade of Cuba. With Lamarr's patent by then expired, the American military simply stole her idea and re-patented it as spread-spectrum communications. Lamarr's concept remained the province of the American Military until it was declassified in 1982, when civil communications companies leapt on the idea like feeding wolves to produce that bane of modern life – the cellphone. Lamarr, who died in modest circumstances in Florida in 2000, never saw a penny.

Starlite

IMAGINE A SUBSTANCE you could simply paint on to anything, from your body to a vehicle, to protect against temperatures in excess of 3000 °C or even the flash of a nuclear blast. Well, you don't have to imagine it because it already exists; trouble is, the inventor died without telling anyone how to make it.

A HAIRBRAINED SCHEME

Maurice Ward (1933–2011), a somewhat truculent and obstreperous hairdresser from Yorkshire, was also an 'inveterate tinkerer' and a true beneficiary of serendipity. Not content with conjuring up his own dyes and tonsorial products, which others of his calling seemed to have travelled for miles to buy, he was also prone to bouts of free-style invention of which the most celebrated offspring was Starlite.

In the early 1980s, a local ICI plant was selling off an old extruding machine and it caught Ward's eye; he picked it up for a song, installed it in a workshop and started playing around with different compounds and mixtures, extruding to his heart's content. Next, Ward heard ICI were after a new plastic for Citroën car bonnets so he set to work but, as he himself said, the results came out of the machine '. . . as scraps. We granulated it; stuck it in bins and forgot about it.' And then the Manchester Airport disaster of 22 August 1985 took place, when a British Airtours flight to Corfu caught fire on take-off, killing all fifty-five on board in forty seconds. The incident sent a genuine shockwave through the North West – one which this author can well remember – but it had a more profound effect on Ward, whose mind turned to finding a material that would not burn. He dragged out the bins of granules from his car-bonnet venture and, at a loss for anything better to do, commandeered his wife's kitchen blender and began whisking them up with solvents to run them through the extruder again. When the first small sheet came out of the machine Ward turned a blowtorch on it and it happily withstood 2,500 °C on the bench. Wanting to see if it would function as a thermal barrier too, Ward put a piece on the back of his hand and hit it with the blowtorch; anticipating a nasty burn, he could not even feel the heat.

FIRED EGGS

In short, Ward had come up with a material that would withstand three times the temperature required to melt diamonds while itself remaining as cool as a cucumber. It would be ideal for anything from impervious fire doors to laser-resistant tanks; hell, you could even paint it on rocket launch pads to keep them cool and safe. No one believed him. Instead of battering down his door and brandishing fat contracts, the scientific and commercial lobbies treated him like a grandstanding charlatan. So, in 1990, Ward decided to promote himself through the well-respected British science-orientated television programme, *Tomorrow's World*.

Presenter Peter McCann opened the show standing behind a bench with an ordinary egg on which he turned a blowtorch; in something under a second the egg shattered. Then the camera panned to another blowtorch held in a clamp with the full blast hard up against another egg that had been coated in Starlite. McCann wandered off to a small jet fighter to discuss the need of fire-retardants in aviation before returning to his egg about four minutes later. The section of the egg under the flame was certainly by then blackened, but he placed that egg, blackened part down, in his left hand and stated it to be barely warm; he then cracked the egg to show it was still raw.

Well, Ward got all the attention he wanted after that, and not just from the commercial sector, of course. The British Atomic

Weapons establishment at Foulness beat a path to his door, hotly followed by other weapons development institutions, and others like Boeing and NASA bringing up the rear.

TESTING TIMES

All tests were conducted under clinical conditions and Starlite passed with flying colours. The boffins at Foulness wanted to go beyond blowtorches and see if Starlite could handle a simulated nuclear blast with equivalent temperatures in the region of 10,000 °C. They did it again and again but only managed to blacken the sample round the edges. Most significantly, the sample was touch-cool after each 'blast'. The Establishment had budgeted for about two hours cooling off between 'zaps' but only had to wait ten minutes; this delay was not due to any reluctance on the part of the Starlite to revisit the inferno, this was the amount of time it took to reset the equipment. The technicians present reckoned they could have zapped the square of Starlite numerous times in close sequence; it was their equipment that could not keep up with the sample. It should be kept in mind here that the most resilient of materials vaporize around the 2,000 °C mark; pure carbon holds out until the 3,500 °C mark and Starlite was shrugging off nearly three times that temperature.

Next to test was the Royal Signals and Radar Establishment at Malvern where Starlite was repeatedly zapped with lasers but was again preserved; laser hits that would have punched through any other material only managed to impose tiny pits at the impact site or, as *International Defence Review* stated in 1993: 'Starlite showed little damage to the surface, merely small pits with the approximate diameter of the beam and with little evidence of melting.' And, Professor Keith Lewis, who supervised the tests at Malvern, stated that Starlite 'had unique properties which appeared to be very different to other forms of thermal barrier material available at the time.' But no one could figure out how Starlite managed to retain its integrity under such onslaught.

A HARD NUT TO CRACK

Only once did Ward let a sample out of his grip when, with great reluctance, he grudgingly entrusted it to a contingent of the SAS who flew it to the White Sands Atomic Testing Site in New Mexico where, again, it remained blithely unmoved by American attempts to nuke it. Back at Foulness, the boffins went the whole hog and subjected their sample to 75 Hiroshimas of energy-blast. Nothing! NASA expressed considerable interest, not least of which was occasioned by the fact that the 75mm thick protective tiles on their Shuttle had a Q-Factor (energy-absorbency factor) of 1.2 as against Starlite's Q-Factor of 2,470 at a thickness of 1mm; some difference! Boeing wanted to Starlite-coat all their planes – especially Air Force One.

But all talks broke down because of Ward's truculence and cussedness – or perhaps he was simply nobody's fool? Either way, he would not bend to the demands of prospective investors. No one – no matter how much they invested – could have a sample in case they tried to back-engineer the material; he would not register a patent and lay bear his secrets; any business arrangement had to be with him holding fifty-one per cent of the control without the other stakeholders being told how he made the stuff. Talks collapsed. Best leave it to Ward's own words on the subject:

A lot of people have been saying I'm a rotten prat, that I'm greedy and I should give it to the world. That's one of the reasons I've tried keeping hold of things. I've said it often enough, I want to give protection, not to cause devastation.

It now seems we shall never know how Starlite was made, because this rather startling piece of serendipity went to the grave with Ward in May 2011. His widow, Mrs Eileen Ward – her name still at the time of writing listed at Companies House as surviving Director of the intriguingly named Starlite Technologies and Stud Limited – has to date kept her own council.

Pykrete

IN THE GERMANY of the mid-1930s Herman Mark (1895–
1992), aka the Father of Polymers, was conducting a series of
experiments to find out why his brainchildren became so brittle
and fragile at low temperatures. He had various samples in a
walk-in freezer in which he had taken the sensible precaution
of salting the floor with sawdust to prevent slips and falls.

During routine clean-ups he and his assistants noticed that
the ice on the floor, compared to that on the shelves and samples,
was more than a little resilient to its removal; it also took an
inordinate time to thaw in the sink. Mark made a mental note
of these factors as nothing more than an interesting puzzle. With
Hitler's star in ascendance and he a Jew, Mark had something
far more important on his mind – escape.

In 1937 he bumped into an Executive of the Canadian International Pulp and Paper Corporation who offered him a research position at the company installation in Hawkesbury, Ontario, where they were trying to make synthetic fibres and polymer yarns out of wood pulp; Mark said he would try to get over there but made no promises. However this chance meeting cemented his resolve to escape to North America.

SKI PASS

Ever inventive and imaginative, Mark came up with a plan to escape with the bulk of his assets while others with Jewish heritage were being stripped of theirs as they quit the country. As quietly and unobtrusively as possible, Mark began acquiring considerable quantities of platinum and, in early 1938, he was ready to make his move.

After paying some hefty bribes, he managed to get his passport released by the Gestapo, along with transit permits to Switzerland for him and his wife to go on a skiing holiday. Dressing and packing in accordance with this cover-story, the Marks stood impassive as they and their meagre baggage was searched at each Nazi checkpoint to make sure they were carrying nothing more than spending money before, with skis strapped to the roof and a Nazi pennant flying cheekily from the bonnet, they crossed into Switzerland and safety. Working on the principle that the best place to hide something is always in plain view, those who rifled their luggage missed the fact that all the coathangers were made of pure platinum.

A CHILLING DISCOVERY

Although he made it to Montreal, Mark had a change of heart and, thanking Canadian Pulp and Paper for the offer, took up instead a position across the border with the Brooklyn Polytechnic, which was anxious to set up its own polymer programme. It was here, in 1943, that he remembered the rogue ice in his freezer back in Vienna and, after conducting a few replicating experiments, published a paper on the incredible strength of ice formed with fourteen per cent sawdust or cotton fibre. No one paid any attention – at least not then.

Across the Atlantic, the British Ministry of Defence was considering the possibility of using levelled-off icebergs as mid-Atlantic airbases and, by fluke, a copy of Mark's ignored paper found its way into the hands of Geoffrey Pyke (1894–1948), a journalist with a taste for the unconventional.

Struggling to make sense of some of the more scientific sections of Mark's paper that dealt with tensile strength and sheer-loads, Geoffrey Pyke turned to mainstream scientist Max Perutz (1914–2002) at Cambridge who, as luck would have it, had been mentored by Mark when he was himself in Vienna. Perutz, who would later supervise Crick and Watson in their identification of the structure of DNA, would later say that had he not known Mark and held him in the regard that he did, he would never have paid one jot of attention to the paper. Basically, Mark's assertion was that ice formed with a fourteen per cent adulteration of sawdust would have an overall strength akin to that of concrete and would endure many times the thaw factors of ordinary ice. It could also be easily fashioned to any profile desired without breaking or its overall strength being compromised.

PYKRETE IN THE SKY

The next stage of this extraordinary story was driven by the boundless energy and enthusiasm of Pyke, who kept up the pressure until even the deafest of Establishment ears listened to his suggestion that, instead of towing natural icebergs into the mid-Atlantic to serve as airbases, this reinforced ice be used to create floating fortresses of up to 2 million tons. These, he enthused, could be maintained in perpetuity by on-board refrigeration plants, with enemy bomb-damage swiftly repaired by seawater and sawdust pumped into any craters in the ice. Simple! His greatest ally in this seeming madness was none other than Lord Louis Mountbatten (1900–79) who requisitioned a large walk-in freezer in London's Smithfield Market for further experiments. Here, Pyke and Perutz soon established that adulteration as low as four per cent sawdust would produce optimum results – ice as strong as concrete that refused to melt within anything like conventional timeframes.

And so was born the outlandish Project Habakkuk, so named after the character in the Bible who admonished all that he would conjure forth 'a work in your days which you will not believe'. Unfortunately, few in the Ministry were as erudite as Pyke and the spelling was scrambled to Habbakuk in all subsequent documentation – but that would be the least of anyone's problems.

Others were quietly questioning the sanity of such a venture but, with Mountbatten behind the project, these people fell in line, agreeing that Pyke was indeed a visionary. To convince Churchill of the value of pykrete, as it was and still is known, Mountbatten records that he nipped down to Chequers, marched unceremoniously into the PM's bathroom and tossed a block of the stuff into the hot bath where he lay smoking one of his trademark cigars. The two men chatted as the small block of pykrete bobbed about unaffected by the hot water and, with Churchill firmly convinced, they agreed that they should unveil pykrete at the impending Quebec Conference, their next scheduled meeting with Roosevelt on 17 August 1943. And present it they did.

THE PLAN BACKFIRES

Ever a boy for his toys, Mountbatten had already tested the strength of pykrete with countless firearms, none of which, including shotguns at close range, had managed to inflict much more than surface scratches, so this he secretly decided would be the most dramatic of unveilings in Quebec. Once the leading lights of the Conference were assembled, Mountbatten took front-and-centre to give a brief talk on Project Habakkuk before whipping out a service revolver and firing it into a block of normal ice, which obediently shattered under the impact. Next, as hoards of security men piled in to the auditorium brandishing their own weapons, Mountbatten fired at a block of pykrete which, oblivious to such intrusion, remained stoically intact as it deflected the bullet to ricochet uncomfortably close to the head of Sir Charles Portal, British Chief of Air Staff, before inflicting a minor wound on the leg of his American opposite number, Fleet Admiral Ernest King.

MELTDOWN

To be fair, it was not just this fiasco that marked the American reticence to stump up $100m to fund the project, it was the fact that they were already committed to the building of modern carriers, which they rightly felt more practicable. But after the Americans had laughed the venture out of court, Mountbatten and Churchill distanced themselves from the project, and from Pyke himself, leaving him to face the deluge of questions that should have been asked long before. If, as everybody knew, ninety per cent of any floating ice-structure's bulk is below the surface, how did he intend to power or manoeuvre these floating fortresses? If everything was to be kept as sub-zero temperatures, where would the crews live? What about the effects of cooking, heating, lighting and a host of other heat-generating activity associated with airbases?

Feeling more than a trifle fragile under this onslaught, on 22 February 1948 Pyke took a fatal overdose, leaving a rambling and recriminatory note to an unforgiving world. Max Perutz, the longest-surviving contributor to the venture would later brand the entire venture an absurd folly, and perhaps it was; perhaps.

AN ANSWER IN SEARCH OF A QUESTION

The popular *MythBusters* television series conducted experiments with pykrete and soon established that potshots with a .45 were water off a duck's back, and that the building of pykrete boats was at least plausible. There is no doubt that the stuff is incredibly strong and cheap to produce – it was even considered for the 1985 construction of the new harbour at Oslo. Perhaps it simply must, for the time being at least, languish as a strange material for which no one has yet found a fit purpose.

A Shipment of Uranium

ON THE 16 July 1945 at White Sands, Alamogordo, New Mexico, Robert Oppenheimer (1904–67) and his colleagues nervously waited to see if their new weapon would work. At 05:29:21, local time, it did. The epicentre of the blast was code-named Ground Zero as a benchmark against which they could measure the devastation radiating outwards at, for example, GZ+1 (mile). This was also the test at which Oppenheimer, staring moodily into the middle-distance, allegedly mused: 'Now I am become death, the destroyer of worlds.' Not only was this a rather loose translation of Chap. 2, v. 33 of the Hindu *Bhagavad Gita*, but it would be another twenty years before Oppenheimer 'remembered' saying it – others, with better memories, were also present. His own brother, Frank, only recalls the more pithy 'Holy fuck, it worked' with Test Director, Kenneth Bainbridge, adding 'Now we are all sons-of-bitches' for good measure. Either way, the important thing about the test as far as we are concerned here is that it further depleted the American stock of enriched uranium. Fine; they knew it worked, but they no longer had enough weapons-grade material left in the kitty to bomb Japan as planned. However, chance was already hard at work solving that problem for them, miles away and on a German submarine in mid-Atlantic.

THE TIME BOMB

It is today decidedly non-pc to say anything of the bombing of Nagasaki and Hiroshima other than that it was a terrible and unwarranted tragedy, but the little-known and less-voiced truth of the matter is that Japan itself was well advanced in its own bomb project. The Japanese Navy had its own programme under Dr Bansaku Arakatsu, this running parallel to that of Dr Yoshio Nishina of the Riken Institute who had started his work as early as 1937.

Fortunately, after their lightening successes in the Pacific, the Japanese Navy lost interest in its own programme at about the same time that Nishina tripped himself up with a scientific accident that can happen to the best of us.

The components in a nuclear warhead have to be slammed together within 1/300th of a second but Nishina seems to have misplaced the decimal point to 1/30th, thus leading the Riken project astray for several months. And this delay seems to have been crucial; a few months before the first bomb hit Hiroshima the Japanese tested their device in the seas off Hungnam in North Korea where they had a secret facility. There are those who claim this test never took place but, if they are right, it is odd in the extreme that, immediately after the bombing of Hiroshima, Russia declared war on Japan with unseemly haste as a special unit was already scorching its way down to Hungnam to round up all the personnel, records and equipment for rapid return to Moscow. At the time, American intelligence confirmed the Russians to be at least twenty years away from developing their own bomb; after their Korean trolley-dash this estimate was revised to three to five years.

THE U-TURN

Finally realizing that the writing was well and truly on the bunker wall, Hitler ordered all the weapons-grade and unrefined uranium to be gathered up for shipment to Japan to allow them to complete their own bomb programme and fight on. This mission was delegated to the U-234, which, under Captain Johann Fehler, put to sea on 15 April 1945 and set course for Japan in the company of two Japanese Lt-Commanders, Hideo Tomanaga and Genzo Shoji. They were in mid-Atlantic when they picked up broadcasts telling them that Germany had surrendered and that all submarines should surface and, flying a black flag, surrender to the nearest Allied port or vessel.

What to do? Fehler had three options: surrender to the British, surrender to the Americans or make a run for some idyllic island retreat and hide from possible reprisals – U-boat commanders were not exactly high on the Allied love-list. The desert island option was soon ruled too fanciful, leaving a choice between Britain and America. Britain was closer to home but had suffered so much from German bombing that Fehler thought they were unlikely to be well received; America might prove more lenient but it was a long way back to Germany from there. Some say that Fehler simply flipped a coin to decide on America but, whether it was that simple or not, he still had the problem of the two Japanese fanatics on board who were both arguing for the completion of the mission as planned. Whether by their own hand or on Fehler's orders, Tomanaga and Shoji were soon dead and, with their bodies hurriedly fired out of Tubes One and Two, it was full speed for America.

U-234 surrending to USS *Sutton*

BUMMED OUT

On 14 May, Fehler surfaced off Newfoundland to surrender to the USS *Sutton*, with each crew untrusting of the other. Everyone was understandably twitchy and decidedly trigger-happy but the only fatality of the surrender was an American seaman who was accidently shot clean up the rectum by the man preceding him down the conning tower ladder of the submarine with his safety-catch off. Taken under escort to the New England coast, the 1,200lb of uranium was swiftly transferred to Oppenheimer who ensured its delivery to its intended destination.

A STROKE OF LUCK

It is difficult to say whether Tsutomu Yamaguchi (1916–2010) was the luckiest or the unluckiest of men. In May 1945 he was one of three draughtsmen sent from his Head Office to the Mitsubishi shipyards at Hiroshima to work on some designs. By 6 August there were just a few last-minute details to tidy away, so all three signed out of the company lodgings and were about to catch their bus, when Yamaguchi realized he had left some paperwork in his room. Telling the others to go on without him, he retrieved the paperwork and caught a later bus. The terminus was perhaps two miles from the shipyard, which left Yamaguchi walking in the open when *Enola Gay* dropped her bomb. Despite his burns and being a 'little deaf in one ear', he managed to make his way through the chaos and catch a train back to his Head Office – in Nagasaki.

At about 11 a.m. on the morning of 9 August, he was standing in his boss's office, explaining what had happened with his boss screaming that he must be mad if he thought a single bomb could do all that. His boss was still ranting when Yamaguchi saw an all-too-familiar flash:

We were both on the floor. The director was shouting, 'Help me! Help me!' I realized at once what had happened, that it was the same thing as in Hiroshima. But I was so angry with the director. I climbed out of the window and got away because I had to help myself.

Leaving his truculent boss to his own devices, Yamaguchi made it home to his wife where, on 15 August, he was sufficiently recovered to sit up and listen to Hirohito's capitulation speech.

Punch-Card Machines

When the young Herman Hollerith (1860–1929) sat watching the actions of a train conductor as he rode home from visiting his girlfriend, no one could have foreseen that he was about to come up with the mechanism that would allow the Nazis to murder so many millions with the chilling efficiency they achieved.

THE ORGAN SCHOLAR

The storing of information on punched card was born of a naughty French boy in the early 1700s being banned from playing out and forced instead to help in his father's organ-building workshop to make up for some transgression or other. At first bored and uncooperative, the young Basile Bouchon suddenly became fascinated by the way air could be channelled this way by the simple expedient of his father closing or 'pulling out all the stops'. Never forgetting this open-and-shut simplicity, the basis of any binary system, in the 1720s Bouchon invented the card-operated loom for which, later, Joseph Marie Charles (1752–1834) would reap all the credit. (Charles is, of course, better known by his nickname of 'Jacquard'.) Under Jacquard, punch-card looms rose to international prominence and became a major inspiration for Charles Babbage (1791–1871) to build his famous analytical machine with his close associate, Augusta Ada Byron (1815–52), daughter of the poet, becoming the world's first computer programmer. But back to the amorous Herman.

PUNCHING ABOVE HIS WEIGHT

Hollerith had learning difficulties and, in all likelihood, dyslexia that took him out of mainstream education to be tutored individually to help him follow his dream of becoming a mining engineer. In 1875, he just scraped a degree from the Columbia University School of Mining. His minimum scores, especially in writing projects and mechanical aptitude, made it hard for him to find a job of his choice, so he took temporary employment in the Washington DC-based Census Office that was ramping up for the 1880 count. It was here that he met and became bewitched by the vivacious Kate Sherman Billings (1866–1933), daughter of John Shaw Billings (1838–1913) who was in charge of the project. Invited back to the Billings home on several occasions, the furthest thing from young Herman's mind was punch-card systems but he had to sit and listen politely to John Billings bemoaning the fact that there was not some kind of machine to take the donkey-work out of the Census which was then an incredibly labour-intensive operation with everything counted and noted by hand.

Keen to impress the parents of the object of his dishonourable desires, Hollerith discussed the matter attentively with her father, soon dismissing the Jacquard loom operating system as too cumbersome. Apart from the fact that one mistake or one alteration to an individual's status would render an entire and ponderous reel completely useless, it was too simplistic a system, like the organs that inspired it, it only worked on an open/shut basis – pick up this thread or don't pick up this thread. What the Census needed was a data-logging system that could cope with each individual person and be able to be quickly loaded with complex and varied information in a mechanical shorthand. Promising to give the matter further thought, Herman took his leave of the Billingses and headed for the last train home.

Cover of the *Scientific American*, August 1890

THAT'S THE TICKET

As he watched the conductor moving through the carriage, checking and punching the tickets, something attracted Hollerith's attention; he noticed that the ticket was always turned to the same position in the man's hand and then punched several times in different locations. When the man arrived at Herman's seat, he asked him to explain what he was doing. The conductor explained that it had all started back in the heyday of the famous train-robber, such as Butch Cassidy and Jessie James, who would frequently put one of the gang on the train as a passenger. Any single man, especially one travelling without baggage, would have his ticket punched in a certain way on issue, before he even boarded the train, to alert the conductor to keep an eye on him. But, as train robberies became a thing of the past and the railways expanded, so too did the punch-code to curtail the increasing number of freeloaders.

This was the part that *really* grabbed Hollerith's attention. These trains travelled great distances with crew and conductors changing at specific points so, to prevent multiple passengers using the same ticket and keep moving round the train to confuse the fresh conductor, the ticket was punched in positions to describe the passenger who first presented it. One position for their sex, another for approximate size and height, another for hair colour, and so on. So, if a six-foot man with red hair showed a ticket punched in such a way as to indicate a short, blonde woman of petite build, there was something seriously wrong. Hollerith saw the light. By the time the 1890 Census rolled around, Hollerith punch-card machines would be used to tabulate the race, colour, creed and other factors of the entire nation and his company, and this first 'garage start-up' of the computer world, would evolve into what is now IBM.

As early as 1933, IBM was in close negotiations with Hitler's evolving regime and would remain in his pocket until the end. To maximize profits from this ill-advised liaison, IBM was always careful not to sell machines to Hitler but to lease them instead on a monthly payment structure and provide the staff to run and maintain them. And so began the mammoth task of cataloguing and categorizing the Jews, Gypsies, homosexuals, Communists, the physically or mentally impaired and all the other 'undesirables' throughout Occupied Europe, flagging them up for either slave labour or shipment out to the death camps on a railway network that was itself run by the same company. Every concentration camp had its own IBM-Hollerith department, which allocated every inmate their own, unique IBM number. The number marked out the holder to be, for example, Polish, so that the camp administration could easily select the next batch for extermination or perhaps for specific tasks.

At the end of the day, countless companies fell over themselves to do business with Hitler and the Nazis – but most just sold them things. IBM stands out as the only company to have entered into the planning and execution of Hitler's darkest vision.

LSD

THE TWISTED TRAIL of LSD started long before Albert Hofmann (1906–2008) took his famous Technicolor bike ride round Basel, Vienna in 1943 – but, unlike Hofmann, those who ingested the precursor of what he would christen his 'problem child' rarely enjoyed the trip.

CATCHER IN THE RYE

Ergot fungus has always infested rye, this problem growing with the expansion into close-order propagation to produce crops of commercial quantities. Ergot infection produces a tiny, purplish spur-like addition to the head of the rye, hence 'ergot' deriving from the early French *argot*, a cock's spur or heel-talon. ('Argot' is an allied term alluding to the vernacular of rough and uncultured rustics.) Unaffected by the milling and baking process, the effects of ergot on those who ate the resulting bread were proportional to the level of infection in the crop; 1 per cent was enough to produce some pretty unpleasant convulsion and religio-spiritual hallucinations, ranging up through gangrene and blackening of the extremities and, finally, death at about 7 per cent infection. With rye being a rural staple, when ergot poisoning visited a community, all hell broke loose. This was far more common in mainland Europe where the warmer weather was more conducive to the fungus gaining foothold, and none more so than in the Italian town of Taranto.

TOURIST CLAPTRAP

As described in the eleventh century, the locals were frequently afflicted with terrible convulsions and religious ravings, claiming to be in the company of angels, saints and, of course, the Virgin Mary. Not understanding the cause and presuming the occupants of the town to be unusually blessed, pilgrims travelled from miles around in the hope of witnessing such divine visitations – and they were never disappointed. Realizing they were on to a winner – pilgrims need accommodation and souvenirs – the locals honed their performances into structured and whirling dances, accompanied by much howling and screaming, before collapsing in a gibbering heap. (No one then understood the internal mechanics of getting 'giddy' which is why that term derives from 'giddig', meaning god-possessed.) Sometimes there were genuine outbreaks of ergot poisoning to lend an additional *je-ne-sais-quoi* to the entertainments but, either way, the locals kept hamming it up until the early seventeenth century by which time everyone, they included, was thoroughly bored with such shenanigans. At its peak, the deception prompted composers to churn out fast and frantic *tarantellas* to mimic the dancing and the local wolf-spider was re-branded as the tarantula, along with the story that the dancing had its origins in some life-saving ritual to avert death from its bite. Unfortunately for this convenient story, as the locals knew full well, the bite of the tarantula, although painful, has never been fatal.

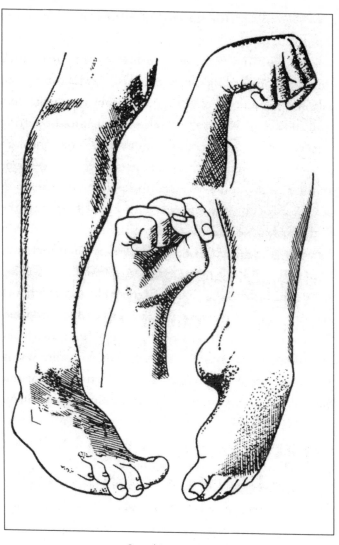

Convulsive ergotism

TRIALS AND ERROR

Ergot poisoning continued to influence history with invading or defending armies being laid low or, on a more localized scale, outbreaks of witch-fever with countless women paying the price for some poisoned yokel claiming to have seen them flying or shape-shifting – ergot-induced hallucinations frequently resulted in one person imagining another to take on a hideous form, which is why some believe ergot to sit at the seat of notions of werewolves. Professor Mary K. Matossian of Yale University has made a study of the influence of ergot poisoning on history and, in her *Poisons of the Past: Molds, Epidemics and History* (1991) she builds a convincing argument by comparing the known outbreaks of ergot poisoning, especially throughout Germany, with the high-spots of witch-fever and mass executions. Some maintain that the infamous Salem Witch Trials were also the result of ergot in the two main accusers, Betty Parris aged nine and Abigail Williams aged eleven, the daughter and niece of the Reverend Parris. There is certainly mention in contemporary record of their enduring convulsions and contortions 'beyond the power of epileptic fits' but other studies of the events conclude they were just a pair of vicious little attention-seekers; either way, they managed to secure the arrest of over 150 and the deaths of twenty-four.

The Salem Witch Trials

BREAKING THE MOULD

It was in later France, the real seat of ergot poisoning, where the mystery would start to unravel. Until the 1630s everyone believed the outbreaks to be just another infection or divine punishment; no one thought of a physical cause until Dr Thuillier, personal physician to Maximilien de Béthune, Duc de Sully (1559–1641), suspected this to be the case. The first thing that struck Thuillier was the fact that outbreaks were far more common in rural areas than in crowded and insanitary urbanity where one would think a disease could more easily rip through a community. But what *really* irked Thuillier was the

197

lack of any discernible pattern, other than the fact that, nine times out of ten, the 'disease' manifested itself among the rural poor. That accepted, how could it be the result of infection or contagion if one member of such a family succumbed, leaving the others unaffected, or an entire family was struck down but not their neighbours? Even those living in complete isolation would suddenly erupt with the symptoms while the rural rich remained largely untouched by the condition. Then fate took a hand.

OLD WIVES' TALES

In 1630 Thuillier decided to see for himself and set out from the comfort of Angers to explore the surrounding countryside but, with evening approaching and nothing having struck him as significant, he told the coachman to head for home – and that is when the wheel snapped off the axle in the U-turn. Realizing that the repairs would take most of the next day, Thuillier checked into the nearest inn which is where he had his eureka moment; rye bread was piled up on platters, freely available as a cheap 'filler' of which the locals and the poorer travellers took full advantage of the fact – but not so the carriage-trade; they all wanted the more expensive white breads. It all fell into place; that is why the rural rich missed out on the fun; they, like the average city-dweller, did not favour the heavy rye breads of the country; they opted for white bread. Deciding to extend his stay, Thuillier was in the rye fields at first light and noted the sporadic appearance of what the farmers called cockspurs in the crop. He spoke to some of the local 'old wives' who informed him that they would collect these tiny spurs to make a preparation to induce labour and inhibit post-natal bleeding – but only in very weak form, anything stronger and the mother was sent screaming round the bend! He had his answer; it all depended on the strength of infection. This also explained why one family could all eat from the same loaf with only one or selected members of that family going down with the symptoms. However, no one believed him.

A FRUITLESS CROP

Others took up the gauntlet thrown down by Thuillier and by the mid-nineteenth century the incidence and lifecycle of the ergot fungus was fully understood, with the aforementioned Albert Hofmann whiling away 1938 tinkering with the fungus in Sandoz Laboratories in Basel. He had isolated lysergic acid from ergot in an attempt to find a purified form of the old midwives' preparation with which he could alleviate migraines and inhibit post-natal bleeding; ergot had by this time been identified as a powerful vasoconstrictor, hence all that medieval gangrene with blackened arms and legs falling off. All his efforts proved fruitless so he shelved the project and turned to other things. But something kept nagging at his mind; had he overlooked something?

A 'TRIP' HOME

The war intervened and it would be 1946 before he again addressed the problem, this time deciding to knock up a tiny amount of lysergic acid diethylamide, to make an easily soluble crystal in the form of a neutral tartrate. Although unaware of it, Hofmann was then standing in the presence of the most powerful hallucinogenic known to man – about 10,000 times more potent than mescaline. He must have accidentally inhaled a minute speck or put a finger to his mouth with the merest trace for, as he would later record:

Last Friday, April 16, 1943, I was forced to interrupt my work in the laboratory in the middle of the afternoon and proceed home, being affected by a remarkable restlessness, combined with a slight dizziness. At home I lay down and sank into a not unpleasant, intoxicated-like condition characterized by an extremely stimulated imagination. In a dreamlike state, with eyes closed (I found the daylight to be unpleasantly glaring), I perceived an uninterrupted stream of fantastic pictures, extraordinary shapes with intense, kaleidoscopic play of colours. After some two hours this condition faded away.

AS HIGH AS THE SKY

Realizing his condition had to be connected to the lab he returned and, deciding that 0.25mg – one quarter of a milligram - would most likely be the smallest amount to have some quantifiable effect, he took that orally and then sat back in anticipation. He did not have long to wait as 0.25mg is about 15 times the amount taken for the average 'trip' today. He started to make notes that mention: 'mild dizziness, restlessness, inability to concentrate, visual disturbance and uncontrollable laughter' but then the writing trails off as Hofmann flipped out on a six-hour 'trip' that left him feeling 'a little tired'. Once recovered, he jotted down what he could remember:

> The last words were written only with great difficulty. I asked my laboratory assistant to accompany me home as I believed that I should have a repetition of the disturbance of the previous Friday. While we were cycling home (a four-mile trip by bicycle, no other vehicle being available because of the war), however, it became clear that the symptoms were much stronger than the first

time. I had great difficulty in speaking coherently and my field of vision swayed before me and was distorted like the reflections in an amusement park mirror. I had the impression of being unable to move from the spot, although my assistant later told me that we had cycled at a good pace . . .

By the time the doctor arrived, the peak of the crisis had already passed. As far as I can remember, the following were the most outstanding symptoms: vertigo, visual disturbances, the faces of those around me appeared as grotesque, coloured masks; marked motoric unrest, alternating with paralysis; an intermittent feeling in the head, limbs, and the entire body, as if they were filled with lead; dry, constricted sensation in the throat; feeling of choking; clear recognition of my condition, in which state I sometimes observed, in the manner of an independent, neutral observer, that I shouted half insanely or babbled incoherent words. Occasionally I felt as if I were out of my body.

Six hours later, Hofmann was still experiencing hallucinations – 'constantly changing colours' and 'fantastic images'. In fact, he had also been trying to climb the walls to escape a concerned neighbour who'd come to see what all the fuss was about; Hofmann saw her as a hideously disfigured witch who flew about the room, trying to eat him. After he came down from the ceiling, his colleagues kept questioning him about the amount he had taken; was he not confused? Could all that have come from one quarter of one milligram? Yes; five grams is enough to produce noticeable effects in about 3,000 people – and it was this strength that took the drug to the attention of those wonderful people who brought you the whole Hippy Movement and the last major outbreak of 'ergot' poisoning.

STRANGE BUT TRUE

Hoffmann's mind-bending discovery came at just the right time for the CIA who, locked into a Manchurian-candidate/ Cold War mindframe, were in the opening stages of one of their most bizarre programmes – MK-Ultra, the search for a good truth-drug and complete 'mind-kontrol' over their own operatives, domestic 'subversives' and the enemy. Inspired by experiments with less potent hallucinogens, such as the mescaline project in Dachau, the CIA 'whitewashed' a load of Nazi doctors and gave them new identities so they could 'join the club' and drive MK-Ultra to its final lunacy. The project's leading light was Dr Sidney Gottlieb (1918–99), real name Joseph Scheider, who with a stutter, a club-foot and a passion for Bavarian folk-dancing was the inspiration for Dr Strangelove; he was also the idiot who came up with all the silly ideas for assassinating Castro with an exploding cigar, etc. While others in the programme were busy producing the famous 1954 cartoon version of Orwell's *Animal Farm*, Gottlieb was ordering vast quantities of LSD from Sandoz Laboratories who, eventually alarmed at the quantities, dropped the gate and refused further supplies. Instead, the CIA 'leaned' heavily on Eli Lily Pharmaceuticals and forced them to break patent so Gottlieb could covertly dose thousands of unwitting Americans with LSD and then try to control what was left of their minds.

A NEVER-ENDING JOURNEY

Many of these guinea pigs were incarcerated or committed to other institutions; some were military personnel and others just random members of the public. In one case, a mental patient in Lexington, Kentucky, was kept under the influence for a staggering 174 days; even when the LSD was withdrawn he stayed on his trip for the rest of his life.

But some came back to roost; in 1959 and throughout most of 1960, the brilliant Harvard student, Ted Kaczynski (b.1942), took part in what he was told would be a stress-control experiment and, with his head so shredded, he resurfaced as the Unabomber who kept the authorities on the hop from 1978 to 1995. And two other rather surprising names crop up on the lists of willing participants of the MK-

Ultra experiments – Allen Ginsberg (1926–97) and Ken Kesey (1935–2001), both leading lights of the so-called Beat Generation and 'bridges' into the Hippy Movement with the latter best remembered today for his *One Flew Over the Cuckoo's Nest* (1962). Most surprising of all is the revelation that the Hippy's guru and LSD advocate, Timothy Leary (1920–96), the man who coined 'Tune in, turn on and drop out', the mantra of the age, was himself nothing but a CIA puppet.

DOSE OFF

By now 'self-medicating' with LSD on a regular basis, Gottlieb moved MK-Ultra to its own downfall with Operation Midnight Climax which was inspired, if that be the right word, by Gottlieb's reading the de-briefing notes of CIA pet-Nazi Walter Schellenberg, SS Brigadefuhrer (retired), a one-time lover of Nazi agent, Coco Chanel. Although Schellenberg's chain of 'tame' brothels were only designed for blackmail and to trap the unwary traitor, the now-twisted mind of Gottlieb decided to replicate the structure and use the tame prostitutes to slip LSD to all their clients. Everything was filmed for Gottlieb who watched avidly while on acid. In 1972 he was 'retired' from the CIA and packed off to remote India to run a leper colony with MK-Ultra's most audacious crime still waiting to be brought into the public domain by a 21st-century journalist.

ACID TEST

On the morning of 16 August 1951, the southern France
backwater of Pont-Saint-Esprit was waking to another day
when the local postman, Leon Armunier, fell off his bike
screaming that he was on fire and under attack from snakes.
He was the first of 250 to lapse into such terrors, with four
deaths, as people threw themselves out of windows to escape
non-existent fires and the like, and two later suicides. There
were immediate assumptions that this was a return of the
medieval horrors of ergot poisoning but this does not jibe
with the victims' conditions – no gangrene or any of the other
classic indicators of ergot poisoning – and then there was the
local doctor who, having eaten nothing that morning, was
rendered speechless and almost catatonic for several hours after
handling a few of the victims. But, most significant of all is
the fact that CIA bio-weapons expert, Frank Olson (1910–53)
is known to have been in the area a short time before the
outbreak and Hofmann was hot-foot from Sandoz to the
town on hearing the news. Hofmann was quick to proclaim
ergot the culprit but this opinion was carefully modified after
his return to Basel and the company who would not long
after cease their dealings with the CIA, who remain the prime
suspect of a crime perpetrated just to see what would happen
in a free release of LSD within an unprepared community.

LUCY IN THE SKY WITH DIAMONDS

This is the Beatles' song that 'everybody knows' to be about LSD; except it wasn't. While righteous indignation was focused on this track, the guardians of rectitude missed the real target – 'Day Tripper'.

John Lennon was inspired to write the song after his son, Julian, then aged five, came home from school with a painting depicting his classmate, Lucy O'Donnell, in a fantasy landscape. Asked what the painting was called he said 'It's Lucy – in the sky with diamonds.' Julian would later state, 'I don't know why I called it that or why it stood out from all my other drawings, but I obviously had an affection for Lucy at that age. I used to show Dad everything I'd built or painted at school, and this one sparked off the idea.' Confirming this in a BBC radio interview in 2007, Lucy herself said, 'I remember Julian and I both doing pictures on a double-sided easel, throwing paint at each other, much to the horror of the classroom attendant . . . Julian had painted a picture and on that particular day his father turned up with the chauffeur to pick him up from school.' As for Lennon himself, he says that the popularly imagined connection never entered his head and he was hardly the kind of chap to shy from such matters. 'Until someone pointed it out I never even thought of it.'

UNLUCKY FOR SOME

As for Olson, he seems to have gone through something of a crisis after the event and been terminally loose-lipped with his opinions and desire to be done with the CIA for good. He had also broken security by twice mentioning the French Experiment, which he had also told his wife was 'a terrible mistake'. Feigning concern, his CIA controllers sent him to New York to see a 'company' shrink but, on his first night in the city, 28 November 1953, he 'fell out of the window' of his thirteenth-floor hotel room. Eventually, the CIA admitted to spiking his drinks with LSD and in the course of researching his book, *A Terrible Mistake* (2010), American investigative journalist Hank Albarelli uncovered a CIA document labelled: 'Re: Pont-Saint-Esprit and F.Olson Files. SO Span/France Operation file, inclusive Olson. Intel files. Hand carry to Belin – tell him to see to it that these are buried'. SO Span would be the codename of a special operation with 'span' likely reference to 'bridge', as in Pont. The Olson family are currently suing their government over the death of their father and some of the remaining survivors of Pont-Saint-Esprit, such as Leon Armunier, now aged ninety, still get acid-flashbacks, but they are coping.

The Frozen Experiment

A COOKERY CLASS in rural Tanzania of 1963 seems an unlikely setting for the accidental birth of an apparent scientific anomaly that still has people arguing, but it was here that Erasto Mpemba set the scientific community on its ear with his assertion that hot liquids freeze quicker than cold ones.

ICE-CREAM HEADACHE

Aged just thirteen, Mpemba was, like the rest of the class, making ice cream but lagging behind through chatting and messing about. He realized that if he did not get a move on there would be no room left in the freezer for his container so he just shoved it in, still hot. To his surprise, his sample was found to be ready to eat first which puzzled him as being counterintuitive to the point of insanity. He asked his teacher how this could be and was told that he must have been confused.

A couple of years later, the school was visited by Professor Denis Osborne of the Physics Department of University College Dar es Salaam and Mpemba, hooted down with ridicule by his peers, stood up to ask, yet again, how this could be so. Osborne, on the other hand, was intrigued and began a series of experiments to see if there was in fact anything in this. And there was, but only under specific circumstances. The blanket assertion that boiling water will freeze faster than cold water, *per se*, does not, if you will pardon the pun, hold water. Besides, Mpemba was working with milk. Water that *has* been boiled and then allowed to cool to the same temperature as an identical but unboiled sample will indeed be the first to freeze as all the bubbles and impurities that would otherwise impede the freezing process have been driven out. But Mpemba and Osborne are indeed right, under certain circumstances, using containers of a specific size filled with identical fluids a certain number of degrees apart in temperature, the hotter of the two will be the first to freeze. And no one is absolutely sure why; there seems to be several factors involved.

THE FREEZE FACTOR

The first of these factors is that the bottom of the hot container will melt any ice on the freezer shelf and thus allow it to sit firm on the elements themselves, with the melted ice re-freezing about the base just for good measure. (Those who have conducted experimental forays into the complexities of the Mpemba Effect and sat their hot sample on a cork mat in the freezer invariably fail to produce the effect.) Secondly there is the evaporation factor from the hot sample that will reduce its mass. Thirdly there is the aforementioned de-gassing of the hot sample and lastly the kind of convection currents that assert themselves during the freezing process. As the fluid at the top of the hot sample cools it becomes heavier than that which lies beneath and so sinks to force up warmer fluid to come in contact with the exposed upper surface. This rise-and-fall exchange activity will be far more pronounced in the hotter sample and so help accelerate the freezing process – but only in specific sample-size; two five-litre poly-pack samples will not play the Mpemba game – I know, I've tried it. The optimum sample seems to be about 70cc of sample fluid in a 100cc container.

THE CREAM OF THE CROP

So, there is certainly something going on here at certain temperature ranges and Mpemba is in good company; the apparent anomaly had been remarked on in the past by such figures as Aristotle, Descartes and Francis Bacon, this latter chap dying of a chest infection he developed while stuffing a chicken with snow in an early frozen-food experiment. But there is also a great deal of false information in circulation as a result of Erasto Mpemba's Effect; there are plenty of websites stating that boiling water will, as a matter of course, freeze quicker than cold, while others casually mention that the effect is well known to ice cream manufacturers who take full advantage of the cost-saving benefits of the Mpemba Effect by freezing their product from boiling. After an hour or so on the phone I could not find one Production Manger in the ice cream business who had either heard of the Mpemba Effect or did not reel back in horror at the thought of shoving hot product into the freezer.

Bibliography

Beavan, Colin, *Fingerprints: Origins of Crime Detection and the Murder Case that Launched Forensic Science* (Hyperion, 2002)

Brown, G. I., *The Big Bang: A History of Explosives* (The History Press, 2005)

Bown, Stephen R., *Scurvy: How a Surgeon, a Mariner, and a Gentleman Solved the Greatest Medical Mystery of the Age of Sail* (St Martin's Griffin, 2005)

Brynner, Rock and Stephens, Trent, *Dark Remedy: The Impact of Thalidomide and its Revival as a Vital Medicine* (Basic Books, 2001)

Fant, Kenne, *Alfred Nobel* (Arcade Press, 2012)

Haiken, Elizabeth, *Venus Envy: A History of Cosmetic Surgery* (Johns Hopkins University Press, 1999)

Infield, Glen B., *Disaster at Bari* (New English Library, 1976)

Jones, Simon, *World War I Gas Warfare Tactics and Equipment* (Osprey Publishing, 2007)

Macmillan, Malcolm, *Odd Kind of Fame: Stories of Phineas Gage* (MIT Press, 2002)

Martinez, Alberto, *Science's Secrets: The Truth about Darwin's Finches, Einstein's Wife and other Myths* (University of Pittsburgh Press, 2011)

Meier, Charles W., *Before the Nukes – the remarkable history of the area of the Nevada Test Site* (Lansing Publications, 2006)

Nichols, Peter, *Evolution's Captain: The Dark Fate of the Man who Sailed Charles Darwin Around the World* (Harper Collins, 2003)

Scalia, Joseph Marl, *Germany's Last Mission to Japan: The Sinister Voyage of U-234* (Chatam Publishing, 2000)

Schulman, Seth, *The Telephone Gambit: Chasing Alexander Graham Bell's Secret* (W.W. Norton & Co, 2008)

Sengoopta, Chandak, *Imprint of the Raj: How Fingerprinting was Born in Colonial India* (Macmillan, 2003)

Whitaker, Robert, *Mad in America: Bad Science, Bad Medicine and the Enduring Mistreatment of the Mentally Ill* (Basic Books, 2010

Picture Credits

Page 51: Library of Congress LC-USZC4-11179

Page 72: Peter Laurie / Hulton Archive / Getty Images

Page 93: Mary Evans / INTERFOTO / Sammlung Rauch

Page 145: Keystone / Hulton Archive / Getty Images

Page 157: Photo courtesy of National Nuclear Security Administration / Nevada Site Office

Page 183: Photo courtesy of U.S. National Archives / Record Group 38

Index

Index

ROTATION PLAN